T0318646

Managing Health in the Genomic Era

Managing Health in the Genomic Era

A Guide to Family Health History
and Disease Risk

Vincent C. Henrich
Center for Biotechnology, Genomics, and Health Research, University of
North Carolina at Greensboro, Greensboro, NC, United States

Lori A. Orlando
Center for Applied Genomics and Precision Medicine, Duke University,
Durham, NC, United States

Brian H. Shirts
Genetics and Solid Tumors Laboratory, University of Washington, Seattle,
WA, United States

ACADEMIC PRESS

An imprint of Elsevier

ELSEVIER

Academic Press is an imprint of Elsevier
125 London Wall, London EC2Y 5AS, United Kingdom
525 B Street, Suite 1650, San Diego, CA 92101, United States
50 Hampshire Street, 5th Floor, Cambridge, MA 02139, United States
The Boulevard, Langford Lane, Kidlington, Oxford OX5 1GB, United Kingdom

Notices
Knowledge and best practice in this field are constantly changing. As new research and
experience broaden our understanding, changes in research methods, professional practices, or
medical treatment may become necessary.

Practitioners and researchers must always rely on their own experience and knowledge in
evaluating and using any information, methods, compounds, or experiments described herein.
In using such information or methods they should be mindful of their own safety and the safety
of others, including parties for whom they have a professional responsibility.

To the fullest extent of the law, neither the Publisher nor the authors, contributors, or editors,
assume any liability for any injury and/or damage to persons or property as a matter of products
liability, negligence or otherwise, or from any use or operation of any methods, products,
instructions, or ideas contained in the material herein.

Library of Congress Cataloging-in-Publication Data
A catalog record for this book is available from the Library of Congress

British Library Cataloguing-in-Publication Data
A catalogue record for this book is available from the British Library

ISBN 978-0-12-816015-2

For information on all Academic Press publications
visit our website at https://www.elsevier.com/books-and-journals

Publisher: Andre Gerhard Wolff
Acquisitions Editor: Peter B. Linsley
Editorial Project Manager: Tracy Tufaga
Production Project Manager: Punithavathy Govindaradjane
Cover Designer: Christian Bilbow

Typeset by SPi Global, India

Working together
to grow libraries in
developing countries

www.elsevier.com • www.bookaid.org

Contents

3. The connection between genetic variation, family health history, and disease risk

Lori A. Orlando, Vincent C. Henrich, and Brian H. Shirts

4. Family-specific genetic variants: Principles, detection, and clinical interpretation

Brian H. Shirts, Vincent C. Henrich, and Lori A. Orlando

5. Genes and cancer: Implications for FHH analysis

Vincent C. Henrich, Lori A. Orlando, and Brian H. Shirts

9. Current and future trends in diagnostics and treatment

Lori A. Orlando, Brian H. Shirts, and Vincent C. Henrich

10. Current and future trends to integrate family health history with clinical programs to improve population health

Brian H. Shirts, Lori A. Orlando, and Vincent C. Henrich

Acknowledgments

The current work was assembled with the input, comments, and suggestions of several individuals who encouraged the efforts of the authors. First and foremost, the authors acknowledge and thank members of 27 families who contributed information for the book. Some of these cases have been presented here, and all of them offered important insights and observations for our team. Dr. Shirts thanks families that participated in the FindMyVariant study, who were inspiration for the examples presented in Chapter 4. Dr. Henrich also thanks the members of his own immediate and extended family who voluntarily contributed historic and health information that was used in Chapter 1.

During the writing and revision phase, several practicing physicians have offered their insights and drawn on their experience to provide us with commentary, suggestions, and encouragement. These reviewers include: Dr. Steven Cox, Dr. Clarence Owen, Dr. Randy Jackson, Dr. Ken Winters, Dr. Kevin Waters, Dr. Roy Orlando, and Dr. Jim Weissman.

The authors also thank members from the Guilford Genomic Medicine Initiative (GGMI) whose work has contributed directly to the development of this book's content including Dr. Elizabeth Hauser, Adam Buchanan, Dr. Geoffrey Ginsburg, Astrid Agbaje, Susan Estabrook Hahn, Carol Christianson, and Karen Potter-Powell along with the physicians who participated in the clinical portion of the project: Dr. Karrar Husain, Dr. Hal Stoneking, and Dr. Jim Oliver. Ginger Tsai at the University of Washington helped create figures used in the book. Lastly, we thank the members of Dr. Orlando's team and Duke University who have built on this work to develop the evidence about the value of family health history and how to optimize technology to support its use in clinical practice: Dr. Elizabeth Hauser, Dr. Geoffrey Ginsburg, and Adam Buchanan, who were also part of GGMI, and Dr. Ryanne Wu, Dr. Rachel Myers, and Teji Rakhra-Burris.

Prologue

The impetus for producing this book is the result of a convergence of views between three coauthors with distinct backgrounds and perspectives: a molecular geneticist and basic researcher who had been extensively involved in establishing a Master's level genetic counseling program 20 years ago and later worked on a primary care program drawing heavily on genetic counseling for chronic diseases (Vincent Henrich), a physician scientist interested in implementing genomic medicine approaches in clinical care (Lori Orlando), and a molecular genetic pathologist whose research had focused on the use of genetic diagnostics for identifying and managing families with a risk for cancer (Brian Shirts). Drs. Henrich and Orlando met when they joined forces in 2007 to work on the Genomedical Connection study—a trial that developed tools for collecting and analyzing family health histories to facilitate the identification of those at risk for hereditary syndromes and initiate appropriate screening and prevention strategies. Over the years, their shared witness and insight about the value of family health history for managing patient health continued to develop and their collaborations continued. Dr. Shirts met Drs. Orlando and Henrich at an NIH sponsored symposium on families and family health history technologies. Dr. Shirts, who pioneered an approach to discovering disease-causing genetic variants through selective testing of near and distant relatives, was a kindred spirit on a mission to increase awareness of the value of family health history. Through our independent and collaborative efforts, it is apparent that even when a connection between genetic variation and disease risk exists, it is often not obvious, and not even attainable. Only a positive family medical history offers evidence of the connection, but finding this relationship requires some appreciation about the functional connection between genes and disease.

The recognition that certain chronic diseases and adverse health events cluster in specific families was suspected for centuries and firmly established in the 20th century as a greater proportion of people lived long into their post-reproductive lives. As the connection between DNA and specific adult diseases was established for genetic variants within specific genes, the momentum to sequence the entire human genome intensified, resulting in the base by base assembly of the 3 billion base pairs that comprise the human genome at the turn of this century. Its completion fed a wave of optimism that the genetic basis of many familial diseases could be diagnosed precisely and treated individually.

Since the completion of the Human Genome Project, significant progress has been made in the diagnosis and treatment of several life-threatening diseases. Nevertheless, the complexity and variability in the genome has been greater than anticipated in the earliest years after the project's completion. Moreover, while genetic variation is a significant risk factor for many chronic diseases, it alone may not be all that's necessary to trigger disease onset. Environment also plays a role and in some cases, is the primary contributor to disease risk. Therein lies the opportunity not only to find who's most vulnerable to a certain disease, but which other factors might be modified enough to maintain good health. This reality simply accentuates the basic observation that most families have some experience with a specific chronic disease risk that involves shared risk factors, even when finding specific ones responsible for the risk cannot be found.

Given the complexity and what still is not known about the genome, a major long-term goal for genomic medicine, the prevention of disease and extension of productive lifespan, remains unrealized. The average lifespan for the populations where genetic testing is available has not appreciably lengthened since the genome project's completion. Major chronic diseases still afflict a substantial proportion of patients and while the treatment for them has reduced some death rates, the cost of post onset treatment is high and the recovery is frequently incomplete. Part of the reason for this difficulty is simply that most of the molecular activities occurring in human cells are still unknown. It may take decades more work to understand them. However, another reason for the spotty progress can be attributed to the difficulty in obtaining a sufficient amount of clinical information to detect and assess a familial disease risk reliably, and then offer interventions that reduce disease risk. Ironically, as progress continues in understanding the genome and the molecular basis of disease, the most effective approach for assessing a patient's health history today and offering strategies to stay healthy will emphasize a complete and accurate family health history.

This book is focused on offering insights concerning the functional connection between a patient's clinical presentation and underlying genomic information even when it is difficult to track a precise genetic association. The goal here is to give the provider an understanding of tools, such as family health history and familial co-segregation analysis that are now available to help patients despite the circuitous and indirect path of events that channel cellular processes toward a greater likelihood of an adverse health event. Throughout the book, clinical cases involving a family heath history are presented to illustrate the application of relevant principles. The medical information is unaltered, though exact family details were changed to ensure anonymity.

Fortunately, the connection between most genetic variants and disease is usually not deterministic. Often the signs and symptoms of a disease are evident in the patient and from their family's health history well before an adverse health event occurs. As will be discussed, the clues that portend an event are often obvious in hindsight simply through awareness of other family member's experiences, and the key therefore is for the provider to recognize, assess and communicate the risks so that the patient is fully aware of the implications and motivated to take action. It is our aspiration that as more cases based on family health history are uncovered and analyzed, medical awareness of its utility can lead to better health outcomes on a broad scale, even as the immense scope of exploring the human genome continues in the decade that lies ahead.

Chapter 1

The growing medical relevance and value of family health history

Vincent C. Henrich, Lori A. Orlando, and Brian H. Shirts

- Common chronic diseases shorten productive lifespan for a large proportion of adults.
- Genetic variation does not account for extended productive lifespan, but it can reduce it.
- Family health history is personal and powerful.
- Family health history is a *robust* predictor of an individual's disease risk and often indicates the urgent need for evidence-based interventions and treatments.
- Family members share millions of DNA variants and a few influence adult disease risk.
- Family health history is a convenient and effective proxy for the effect of genetic variants that increase disease risk.
- Inherited alterations of gene sequence or activity can predispose adult disease onset.
- Familial disease risk is a product of both heredity and shared environment.
- Collection and analysis of useful family health histories pose challenges for healthcare.
- The utility of family health history will depend upon new tools and updated models of clinical patient flow.

The general premise of this book is that adult onset diseases, health events, and life-limiting conditions which afflict a previously healthy individual occur neither randomly nor unpredictably. By knowing about the medical history of the patient and family members who are genetically similar to them, it is possible to assign a personal disease risk. Further and more importantly, they and their family members can reduce their personal risk for a specific disease or condition by undertaking specific interventions and treatments starting years

Managing Health in the Genomic Era. https://doi.org/10.1016/B978-0-12-816015-2.00001-8

and even decades before the occurrence of a health event. For the provider, assessing a patient's personal disease risk depends substantially on an ongoing evaluation of vital signs, clinical test results, personal medical history, and family health history. For the patient, sustained good health depends on following an intervention program before health status is impaired. If this approach is to be effective, individuals will consider the health conditions seen in relatives and ancestors, including parents and siblings (first degree relatives) and more distant relatives, including aunts, uncles, half-siblings, and grandparents (second degree relatives). By taking family health history into consideration, they will make medically beneficial decisions with the goal of maintaining a healthy life.

The guiding purpose of this book is to approach the challenge of maintaining health with the recognition that inherent predispositions, as evidenced by family health history, are known to exert a significant effect on adult disease risk in at least a substantial fraction of people, and that it captures the complex and multifaceted relationship between genes, cell biology, and developing disease. Basic genetic principles will be described periodically throughout the book, with special attention given to their direct relevance for clinical practice. These principles and their limitations will be introduced and reinforced through real examples involving families. References and website links at the back of this book will provide additional information for readers interested in learning more about the genetic and genomic concepts discussed in the chapters.

As will be explored throughout this book, family health history can be used to identify familial disease risks, anticipate the age of onset, evaluate environmental influences and triggers, guide testing for potential genetic variants that are predisposing to specific diseases, and detect the early signs and symptoms that mark the onset of a change in health status. An above average disease risk for a given individual also has substantial medical implications for other family members. If a middle-aged adult is susceptible to an adult onset disease, then his/her relatives, including children, may also share that risk and may benefit from further follow-up. As health providers realize, the onset of many adult diseases and health events are actually endpoints that follow a period of, frequently, asymptomatic deteriorating health that often, but not always, is detectable for those with a discerning eye. Family health history is an important clue to look for those early signs. Ideally, when a disease vulnerability is recognized, and effective interventions are offered while a person is young and asymptomatic, the ability to prevent the cost and anguish of adult disease and thereby extending a person's productive lifespan is enhanced. This book will focus exclusively on adult onset diseases and health events, while recognizing that the cellular basis for these conditions and diseases are rooted in mechanistic failures at the molecular level, and the cause-effect relationship tends simply to be more direct and obvious for early age diseases.

Common chronic diseases shorten productive lifespan for a large proportion of adults

The foregoing discussion will concentrate on assessing a healthy individual's susceptibility to affliction by common adult onset diseases and identifying effective interventions and treatments which forestall their onset. The current NIH director, Dr. Francis Collins, has wryly observed: the average number of deaths per person is 1. Therefore, a more realistic goal for disease prevention is to extend the productive lifespan of individuals, during which the pursuit of life activities is unimpaired by health limitations. When evaluating the potential impact of medical practice directed toward maintaining wellness it is worth asking: What is the prevalence of adult onset diseases and how deleterious is the impact of adult onset conditions and diseases on human health and longevity? More to the point, is it worth the investment to manage risk before an anticipated medical condition appears? How many lives are actually shortened or impaired by adult onset chronic diseases? And by how much? According to the Center for Disease Control's 2015 data (Table 1), deaths attributable to several common chronic diseases increase by five to fifteen-fold between the ages of 45 and 65.[1] Given that the average life expectancy is roughly 75 years, an even more pressing question is: Are there evidence-based interventions to prevent these common adult onset diseases that can measurably *and economically* reduce incidence in those who would otherwise reach their average life expectancy?

The *potential impact* of adult disease prevention can be inferred from the current distribution of age groups within the United States. It's a foregone conclusion that life expectancy has generally increased over the last 50 years, though life expectancy is highly variable between racial, ethnic, gender, geographic, and socioeconomic groups. Notwithstanding these concerns, the overall mean life expectancy for today's healthy 50-year-old is 25–30 years according to data compiled by the Center for Disease Control.[2] However, mean life expectancy obscures a morbid statistic— about 35% of today's half-century celebrants will never experience their 75th birthday. And of those who do make it, many will suffer from any one of numerous disorders (dementia, autoimmune diseases, cancer, heart disease, infections, etc.), or be disabled, or deal with difficult rehabilitations and treatments that compromise their activities. Closer inspection of the distribution of five-year age groups confirms that a significant proportion of adults disappear between 50 and 75 years, and that the decline is even greater in some subgroups, notably African-American males.[3] Whereas every 5-year age group in the United States had well over 20 million persons in 2015 (23 million were aged 50–54), the number was below 20 million in the 65–69 age group, and 13 million in the 70–74 age group (Fig. 1). In 2017, the mortality rates for Alzheimer's disease, diabetes, lower respiratory disease, and stroke increased, even as the rates for heart disease remained the same and those for cancer declined slightly. Both of the 65–69 and 70–74 age groups belong to the Baby Boomer cohort, the largest in the history of the United States.[4]

TABLE 1 Leading causes of death and number of deaths by age group, 2015.[1]

Rank	25–34	35–44	45–54	55–64	65+	Total
				Age groups		
1	Unintentional injury 19,795	Unintentional injury 17,818	Malignant neoplasms 43,054	Malignant neoplasms 116,122	Heart disease 507,138	Heart disease 633,842
2	Suicide 6947	Malignant neoplasms 10,909	Heart disease 34,248	Heart disease 76,872	Malignant neoplasms 419,389	Malignant neoplasms 595,930
3	Homicide 4863	Heart disease 10,387	Unintentional injury 21,499	Unintentional injury 19,488	Chronic low respiratory disease 131,804	Chronic low respiratory disease 155,041
4	Malignant neoplasms 3704	Suicide 6936	Liver disease 8874	Chronic low respiratory disease 17,457	Cerebrovascular 120,156	Unintentional injury 146,571
5	Heart disease 3522	Homicide 2895	Suicide 8751	Diabetes mellitus 14,166	Alzheimer's disease 109,495	Cerebrovascular 140,323
6	Liver disease 844	Liver disease 2861	Diabetes mellitus 6212	Liver disease 13,278	Diabetes mellitus 56,142	Alzheimer's disease 110,561
7	Diabetes mellitus 798	Diabetes mellitus 1986	Cerebrovascular 5307	Cerebrovascular 12,116	Unintentional injury 51,395	Diabetes mellitus 79,535
8	Cerebrovascular 567	Cerebrovascular 1788	Chronic low respiratory disease 4345	Suicide 7739	Influenza & pneumonia 48,774	Influenza & pneumonia 57,062
9	HIV 529	HIV 1055	Septicemia 2542	Septicemia 5774	Nephritis 41,258	Nephritis 49,959
10	Congenital anomalies 443	Septicemia 829	Nephritis 2124	Nephritis 5452	Septicemia 30,817	Suicide 44,193

Data Source: National Vital Statistics System, National Center for Health Statistics, CDC. Produced by: National Center for Injury Prevention and Control, CDC using WISQARS.

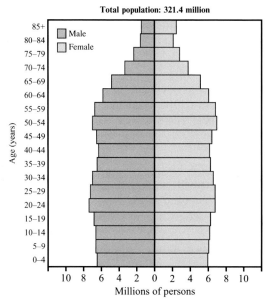

FIG. 1   Distribution of US population by age group and gender, 2015.[2] Population scale indicates millions of people.

Genetic variation does not account for extended productive lifespan, but it can reduce it

It is tempting to jump to the conclusion that longevity is genetically based—some of us are lucky and will live long lives, while others will not, and we can exert only a marginal influence on that truth. It's heartening to know that in reality, population studies have failed to show a strong connection between genetics and longevity, though the familial connection has been understudied.[5,6] Right now, only 25% of the variability in longevity can be attributed to genetic factors. Further, even among nonagenarians, extended longevity is not simply the consequence of good genes, but rather healthy and independent living.[7] Among elders, those who are still independent tend to remain independent. In other words, a longer lifespan is not typically the consequence of a lengthened period of seriously compromised health status. These epidemiological studies, however, must be interpreted carefully. For some individuals maximizing longevity may require a greater level of self-management than it does for others, because some must deal with inherent risks. In fact, most of us have some inherent risk for a chronic disease that could shorten our productive lifespan though the chances of leading an extended and healthy life exists despite them.

Family health history is personal and powerful

Each one of us has a family medical history. Even those coming from families with a long and healthy lifespan on both the paternal and maternal side, the question might be: what vulnerabilities have been forestalled or avoided by our relatives and ancestors? The nature of family-based adult disease risks will be explored, along with strategies for analyzing and interpreting family health history information. We will postulate that it is critical to understand the patterns that precede and potentially foretell the development of an unhealthy condition if the goal is to maximize the benefits of their family health history.

For several years, Dr. Lori Orlando and I worked with a team of researchers and providers in a local health care system to implement tools that assist in the collection of family health history and offer preventive interventions that led to many of the insights and topics compiled here. The program also heightened my awareness of my own family health history, notably that my aunt had experienced a sudden stroke in her 60s without any indications of a pending problem. I also had a vague notion that I had ancestors who had suffered from strokes. My knowledge proved to be woefully incomplete when I was unexpectedly afflicted with a stroke myself.

My role transition from Education Director in our genomic medicine initiative to a patient dealing with my family health history overtook me over a 2-day period in August 2013. A work colleague came by my home on a Monday afternoon after I had missed a major meeting at the University. I had suddenly lost consciousness and collapsed, sometime between Sunday evening and early Monday near the front door of my home. I regained consciousness only after he had knocked on the door for several minutes. Suddenly awakened, I asked my colleague to call 911—while noticing that I had no sensation or movement on my left side. I remember nothing about my loss of consciousness or how I ended up where I was found. As I regained consciousness and awaited the ambulance arrival, I remembered feeling as if I had been contracting the flu over the weekend before. I had canceled my regular Saturday tennis matches, and decided not to attend a University function on Sunday because I felt feverish and lightheaded with a dull headache. My skin had felt clammy and cold. The empty water bottles next to my bed confirmed that I had developed an unquenchable thirst. My rescue, owing to the fact that my colleague acted on his concern to check up on me, began my own education as a patient. What I have learned since that day has helped me appreciate the enormous potential for using family health history to maintain good health, but the experience also revealed numerous gaps and needs that must be addressed before the full utility of family health history as a part of medicine will be realized.

In the weeks prior to my stroke I had experienced a high level of stress over a potentially explosive personnel matter that I inadvertently discovered at my University. The drama had persisted for months and brought one unpleasant surprise after another with no evident path to a constructive resolution. I had also found that I could not sleep restfully during a two to three-week period that preceded my stroke. I correctly attributed my sleep problems to my work stress, but I failed to consider that I was developing severe hypertension. I never had shown hypertensive problems before and I was confident enough about it that I felt no anxiety whenever my blood pressure was measured. That self-delusion was busted up when I read on my patient record that my blood pressure was 240/110 when I was admitted to the hospital. I met a brain researcher years later who asked about the symptoms I had experienced prior to my stroke. She noted that my suddenly erratic sleep patterns, my feverish sensations and my unslakable thirst just before my collapse were consistent with an impairment of functions in the thalamus, which was the site of my stroke.

My own family health history had seemed relatively unremarkable to me. As noted, one of my paternal side aunts had suffered a sudden stroke in her 60s but survived and my paternal grandfather had died at age 66 during an exploratory surgical operation. Actually, some in my family speculated that it was my grandfather's pipe smoking and scotch drinking that caused his problems, though neither had seemed excessive to me. His wife, my grandmother, died a few years later after she was struck by a sudden heart attack. There is no verifiable family health history information about the maternal side of my family other than my maternal grandmother lived to the age of 86, though I was told that my maternal grandfather and his own father had collapsed and died mysteriously before age 60. Both were described as heavy smokers and alcohol consumers. My general awareness about my relatives' bouts with stroke or sudden collapses had prompted me some years earlier to lose weight and alter my diet. The dietary transition marked my reluctant acceptance that no amount of exercise could offset the effects of a high fat diet on my BMI. I had wistfully accepted that my diet would be overtaken by grains and green vegetables for the rest of my life if I was to remain trim enough to play tennis. I also began taking aspirin daily because I had read that it reduces the risk of strokes and heart attacks.

Early in my hospitalization, I began to become aware of my own misunderstanding about strokes and my family's vulnerability, when I asked a neurologist if I should resume taking aspirin to reduce my risk of another stroke and he curtly responded, "You didn't have *that* kind of stroke." I had never known that 85% of strokes are ischemic, but that I had been diagnosed, according to my patient record, with a subarachnoid hemorrhagic stroke. Hemorrhagic strokes often arise from an aneurysm rupture. Those affecting the thalamus, as mine did, and other deep brain structures, comprise about 3% of all strokes, and are frequently lethal, tend to occur at an earlier age than ischemic strokes, and sometimes cause a rapid loss of consciousness. The risk for this type of stroke is 3 to 4-fold higher if one has a positive family history.[8] Ironically, my daily habit of taking small aspirin doses to prevent stroke may actually have increased my risk. Only as I processed my physician's response to my question about aspirin did I fully recognize that my paternal grandfather's brain aneurysm and his daughter's stroke likely involved the same vulnerability that had affected me. I also began to realize that while there have been many well-intentioned efforts to improve public knowledge about stroke risk, it can be misleading and even incorrect to apply the same general term to a health event that could arise from a completely disparate failure. However, health events can often afflict different family members in similar ways, providing clues to their source. As my case shows, failing to recognize these patterns can negate the value of collecting one's family health history. I had failed to appreciate that *family health history is a starting point for diagnosis*, not a complete piece of information by itself. The problem is compounded by the simple reality that the cause of death assigned to deceased ancestors often is not accurate, and may involve comorbidities arising after disease onset or a significant health event.

My curiosity about my family health history was stoked further by an accidental discovery about a year after I had my stroke. I came upon my paternal great great grandfather's passport from Germany to the United States in 1866, while searching for some estate paperwork at my parent's longtime residence. This was one of numerous documents about our family that included obituaries, letters, and death certificates which had been collected by my resourceful uncle over 35 years ago. Most people would never make such extraordinary efforts to learn about their own family's health history, but perhaps my findings will illustrate how much information is harbored in our family's past if we can find it. As I read his findings, I was grateful for my uncle's genealogical work long ago. I learned that my great grandfather also had fallen unconscious from a sudden stroke during the height of the Depression in 1931. He suffered a second stroke and fell unconscious again as he left the hospital a few weeks later, recovered, and passed away after suffering a third sudden stroke in 1936. The first stroke had apparently occurred shortly after he had learned that a trusted and longtime employee at his shoe business had embezzled a large sum of money from him. My great grandfather was slim and athletically active. I also discovered a letter from a cousin's wife in Germany that requested financial help from my great grandfather in the aftermath of World War I. The letter explains that one of my great grandfather's cousins had been disabled by a stroke, suggesting again that my extended family's susceptibility could go as far back as the mid-1800s. That inference was later validated when I learned from a death certificate that my great great grandfather, the one whose passport I had found, had died from apoplexia, the historical term for a cerebral hemorrhage.

The stressful trigger for my great grandfather's stroke prompted me to ask my aunt if she remembered anything unusual that happened just prior to her stroke. She emphatically answered that her stroke had occurred within days after she and her husband learned that he had been diagnosed with leukemia and was given only a few months to live. She also added that her stroke came without warning as she cooked dinner in the kitchen one night. One moment she remembered preparing dinner, and her next recollection was lying immobile in a hospital bed. The sudden occurrence of all these strokes, without many of the symptoms typically ascribed to their onset, has also left me reflecting on the familial nature of symptoms associated with health events and disease onset that will be explored in later chapters. Conversely, these similarities in onset and presentation might also reflect shared mechanistic vulnerabilities at the molecular and cellular level.

As I have discussed health matters with my extended family, it has become apparent that some members are susceptible to bouts of severe hypertension, and others are not. My afflicted aunt's sister (another of my aunts) never developed serious problems with hypertension, nor have any of her six children, now 55–75, at the time of this book's writing. While the evidence suggests that our family strokes are not ischemic and do not affect cognitive functions (e.g., members with left brain strokes were always able to speak normally during recovery if they survived), this level of diagnosis was not available for my ancestors.

Another perplexing issue centers on the circumstances that preceded our family members' strokes. Many of my relatives have led active lives which regularly required hectic work schedules balancing personal and work-related activities that might be described as "stressful." Nevertheless, at least three of the four strokes that have afflicted a family member were tied to an unusual and unexpected source of stress, in addition to those hectic activities normally encountered by most of our family members. My grandfather's stroke occurred less than a year following his retirement. He had been admitted to the hospital after complaining of dizziness and blurred vision over a period of months and in one instance, had been involved in an automobile accident because of it. A brain aneurysm ruptured during an exploratory surgical operation. I only learned that detail from my aunt 50 years after it happened. My own father would logically be viewed as potentially vulnerable to this type of stroke, but he has never experienced any symptoms suggestive of an impending stroke. He seems to be an escaper, who has never had such an event despite his father, sister, and son having experienced a hemorrhage apparently caused by vascular ruptures.

I now view my father's good health as a factor that confused me when making my own risk self-assessment. I have gained a much greater appreciation for the relevance of *incomplete penetrance*—a genetic term—and what it means as genomic information becomes integrated into medical care: Not everyone who carries a genetic variant associated with a particular chronic disease will actually be afflicted by it. It will be a recurring concept in the discussions presented later in this book. In my family's case, the concept has provoked some thoughtful reflection. Is it possible that my mother's insistent supervision of my father's medication regimen, diet, sleep, and lifestyle has protected him from the circumstances that could lead to stroke? Did he simply manage to avoid an unusually stressful situation that might have provoked one or more hypertensive spikes that would lead to a stroke? Stress is an environmental factor, and like many chronic conditions, the combination of inherent predisposition and environmental trigger can lead to a "tipping point" that in turn, brings about a health event or chronic condition. My neurologist mused skeptically when I explained that stress might have been a causal factor for my stroke. He asked rhetorically, "What's stress?" Actually, stress has been defined as a stimulation of steroidogenesis in the adrenocorticotropic axis, but this definition encompasses many distinct and interacting components which remain incompletely described.[9] More generally, what aspects of any environmental factor, stress in my case, are truly the ones that exacerbate disease risk? By what mechanisms does an environmental factor change cell biology and open a path to disease onset? This principle will be examined more thoroughly later in the book; but in this respect, genetic testing can contribute not only to predicting the likelihood of disease onset, but also to identify the cellular mechanisms that are modified in a genetic variant carrier that, in turn, elicit disease onset. This is more complex than it seems. Even for a specific condition, whether a specific cancer type, cardiovascular disease, or metabolic disease (to name a few), there are many regulatory steps, any of which might be compromised and lead to a similar clinical presentation. Generally, family health history reports on the consequence of intersecting and interactive genetic, cellular, and environmental factors shared by family members which have resulted in a positive family health history. Put another way, the physiological basis of a specific disease may vary widely among different families, but the chances are higher that the mechanisms responsible for a specific familial disease are shared between affected relatives. This interplay of risk factors can conceivably be estimated to produce disease risk calculations and it is the rationale for developing risk estimates such as those generated by Tyrer Cuzick or Gail models for breast cancer.[10–12]

Family health history is a robust predictor of an individual's disease risk and often indicates the urgent need for evidence-based interventions and treatments

Aided by my uncle's efforts, I compiled a family pedigree and some family health history information that spans through five generations now (Fig. 2). I still do not know what, if any, genetic variation underlies my family's encounters with stroke, but I have developed a greater sense of urgency about my day to day health status. Most importantly, I discovered how "stress" manifests itself in my body. I learned that adept cognitive compartmentalization of stressful encounters is different than reducing a stress response. This recognition has led me to seek a variety of means to allocate my time and limit my exposure to stressful activities. Second, I learned that my blood pressure seems highly volatile and that diet and exercise is not effective by itself for ensuring that my blood pressure remains stable and I am currently taking three anti-hypertensive medications. After a discussion with me about my stroke, one of my cousins who is active and fit noticed that he was experiencing transient and severe bouts of hypertension during a stressful period in his life. He sought his physician's advice and was also prescribed a hypertensive medication. He also stopped taking aspirin and he is in excellent physical health as he approaches age 70. Both of my siblings now monitor their blood pressure regularly and take medications that would prevent spikes of hypertension. My aunt who experienced her stroke over 20 years ago is healthy, intellectually engaged, well-spoken, and also consciously attuned to her health status. For all of the family members with whom

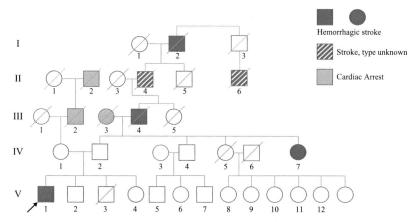

FIG. 2 The Henrich family's pedigree and family health history of stroke incidence. *Squares* indicate males and *circles* indicate females. Generations are indicated as *rows* with members by order of birth. Matings are indicated by *horizontal lines* between parents and *vertical lines* connect to offspring. *Diagonal lines* indicate deceased individual. *Diagonal lines* indicate deceased family members.

I communicate, adhering to medication, diet, and lifestyle recommendations are now viewed as possibly a matter of life and death.

As much as I would like to know, it would be expensive and difficult to find a familial variant responsible for our family's stroke risk. The difficulties are largely technical and also apply to the search for most predisposing variants affecting families. The more salient issue for our family is: what action could be taken to reduce our individual risk for experiencing this type of stroke, if we knew about a predisposing genetic variant? Diagnostically, it would be possible to identify familial carriers of the variant, and thus those who are most vulnerable to a hemorrhagic stroke. Carriers might then be monitored more frequently for evidence of hypertension, and might be recommended for periodic MRI scans to detect brain aneurysms. Note that by differentiating carriers within a family, relatively expensive tests, such as MRIs, could be targeted to those members for whom a cost-effective benefit is apparent. In addition, if an aneurysm that posed a high risk of rupture was identified, surgical clipping or "coiling" procedures could be employed to tie off circulation to the aneurysm. Family health history, appropriate diagnostic testing, follow-up monitoring, and individually targeted interventions and treatments describe a set of guiding principles for personalized and precise medical practice. However, it is noteworthy that even in the absence of a genetic test that distinguishes high risk family members from those with an average risk, my family health history provided sufficient information for my relatives to take appropriate action in consultation with their providers.

Family members share millions of DNA variants and a few increase adult disease onset risk

The collection of DNA (deoxyribonucleic acid), and the genes therein, that each person carries is referred to as the **genome**. It is now well-established that our understanding of human biology, and more specifically, human health, requires an understanding about the structure and function of the over 22,000 genes which lie at specific locations along our 23 pairs of chromosomes. Each chromosome is a highly structured complex of DNA and proteins, and the entire 46-chromosome collection inherited from one's parents lies in each of the 5 trillion nucleus-containing cells in our body (each of us has another 20 trillion or so enucleated red blood cells). One of each chromosome pair in a person's cells was inherited from their mother and the other from their father. Our individual features, tendencies, capacities, susceptibilities and traits are influenced by what we inherited as a result of the chromosomal shuffle that occurred in each parent's germ line (i.e., the genes in the cells (sperm and ova) that are involved in reproduction). In fact, exactly *one copy* of each of those 22,000 genes came from each parent. The normal exception to this rule is that males inherit one X chromosome from their mother (along with a Y chromosome from the father) and therefore, carry only one copy of the genes that reside on the X-chromosome. There are also other rare situations where exceptions to these rules happen, but these do not occur normally and will not be considered here. Those 22,000 genes actually comprise a small portion of the roughly 3 billion DNA base pairs that we obtain from each parent (each referred to as a **haplogenome**). In turn, those two haplogenomic copies (3 billion base pairs each) are replicated and packaged into almost every cell in our body. Roughly 98% of the DNA in the human genome is located within intergenic regions (the sections of DNA that are between genes (protein coding sections of DNA)). The functional importance of intergenic regions is mostly unknown, though there are instances when altering the sequence within these vast regions of DNA sequence seems to alter gene function, but the mechanistic basis for this connection involves a discussion beyond the scope of this writing. Each one of us carries several million DNA variants. About half came from each of our parents and we share some of those variants with our siblings, and smaller fractions with more distant relatives. Each one of us adds a few new variants to the family collection, too.

The underlying point here is that a gene associated with disease is not something to be extracted, nor is it the sole determinant of a person's disease risk. Rather, genes like *BRCA1* and *BRCA2* normally direct the production of a protein that in turn, performs one or more cellular functions. This explanation leads to the most fundamental genetic concept relating to family health history—which we will designate as a "***DNA variant***," that is, an alteration in the DNA sequence. The entire DNA sequence of the human genome is defined by the order of four deoxyribonucleotides, a molecule derived from the ordered combination of a nitrogen base, a deoxyribose sugar, and one or more phosphate ions. The chemical nomenclature is more complex than described here, and for the purposes of our discussion, A, C, G, and T correspond to the four nucleotides, and their order along a DNA strand encodes the information that specifies a gene's DNA sequence. Each A is paired with a T, while each C is paired with a G to form a double helix comprised of two complementary DNA strands. A thorough discussion of gene activity and its regulation would result in a multivolume encyclopedia. Rather than wade into the topic, this book will mostly concentrate on the medical implications of genetics in a family. The general takeaways from this discussion are that the relationship between genes and disease involves changes in the DNA sequence (variants) which sometimes (but not always) alter the activity of a gene or the function of a protein specified by the gene's sequence. Alterations of the DNA nucleotide sequence that affect a gene's activity or the RNA sequence transcribed from the gene will be collectively described here as **"*genetic variants*"** simply to emphasize that the change in sequence occurs within the small fraction of the genome that lies within the 22,000 genes. A given gene resides at a specific and invariant location (known as a "gene locus") on one of chromosomes. The locale is so precise that the human genome database typically assigns a number for each of the 3 billion bases in the genome allowing for direct identification of specific sequence variants that deviate from the average human's at that point. We know the "average" human's genome based on the work done by the Human Genome Project and other studies, and refer to it as the consensus reference human genome sequence, which is described in more detail in later chapters. The reference genome delineates genes by the numbered bases residing within it.

The clinically relevant point that emerges is that this orchestra of gene activity varies to some degree in every one of us not only because our environments are different, but also because every person carries millions of DNA variants and a few of these will affect cellular processes. These will be a focus of periodic inquiry, especially as it relates to familial disease risk. This volume will not extensively delve into molecular genetic mechanisms, but it will become apparent that variations in the events going on at this level can affect an individual's health status, which in turn, has considerable bearing on diagnosis and treatment. Moreover, the people most likely to share these variable mechanisms are their relatives.

As a career geneticist, I've frequently found myself fielding questions about familial health problems. On several occasions, people expressed concerns about carrying a gene that causes disease, such as a "cancer gene, *BRCA1*" or an "Alzheimer's gene." Many have heard about *BRCA1* or *BRCA2*. Often, they are taken aback or confused when I respond that I carry two copies of *BRCA1* and two copies of *BRCA2*, too. I received one copy of those genes from each of my parents. In fact, everyone carries two copies of each gene (with the aforementioned exception concerning genes on the single X-chromosome that males normally carry). Each person receives one copy of every gene from each parent. In turn, each parent will randomly pass along one of their two gene copies to each germline (sex) cell. This ***Law of Random Segregation*** for the transgenerational passage of one of the two gene copies carried by a parent was originally proposed by Gregor Mendel from his studies of garden pea traits and with a few unusual exceptions, this law of genetics has proven true for the inter-generational passage of genes (and gene variants) in all eukaryotic organisms. Many students have asked me over the years why they must learn about pea heredity when their goal is going to medical school. The short answer is: The Law of Random Segregation is applied whenever a patient and her/his relatives are given a genetic test. As will be seen in cases later on, random segregation is the mechanism by which some family members are genetically vulnerable to a disease, whereas others are not. Moreover, it is common to find that these dichotomous levels of disease risk within a family indicate something genetic may be involved, even if it is impossible to track down a specific variant.

It is intuitively apparent that the collection of variants in each person's ge-nome, along with the random shuffling process that sequesters a haplo-genome into each male and female germline cell, could lead to a myriad of gene variant combinations in both the sperm and egg that eventually combine to produce a baby. A relative handful of all these variants have been shown to influence adult disease risk at some discernible level. For every one of the 6 billion plus bases that comprise the genome (that is, two haplo-genomes of 3 billion base pairs each), there typically is a most common nucleotide at each position which is ancestral to the human species. Here, we will usually refer to this ancestral base (A, C, G, or T) as the *normal variant* or *ancestral variant*. It is worth noting that "normal" is not completely synonymous with "optimal," especially when one contemplates the environmental challenges of humans living in the 21st century compared with humans who lived prehistorically. As already noted, any change in the DNA sequence within one's genome is in fact, an *alternative variant*, often called the *minor variant*, since almost always this variant is less frequent than the ancestral variant. However, the true measure of a given vari-ant's clinical impact depends on what variant *phenotype* (e.g., a trait, condition, symptom, or clinical measurement) is seen in individuals who carry the alterna-tive variant. The variant composition at a given gene (or genes) is referred to as the *genotype*. The *genotype-phenotype* relationship is a fundamental approach for assessing an individual's disease risk, though this relationship can often be inferred by careful observation of families who display a suggestive pattern of disease incidence. Later, the extent of human genetic variation in the popula-tion will be discussed further in order to introduce some essential principles to consider as it pertains to individuals and their relatives.

Family health history is a convenient and effective proxy for the effect of genetic variants that increase disease risk

The early chapters of this book will examine the connection between family health history, genomics and cell biology, using several cases to illustrate the concepts involved and their relevance. The purpose of this discussion is less centered on a comprehensive analysis of the scientific principles at play and more centered on what these principles mean when assessing an individual's disease risk with the goal of maintaining good health.

The rationale for using family health history when assessing disease risk is that among all the people who ever existed on Earth, the ones most genetically similar to each one of us are the members of our own family, especially our parents, our full siblings, and our children, that is, our first degree relatives. The consequence of knowing about familial disease occurrences is that it is possible to understand that an inherent vulnerability exists within a family without going through the usually difficult process of finding the variant responsible for it. As we've seen, all of us carry many variants, but even if each one of us obtained our entire DNA sequence we could say very little about what this means for our future health status. We will revisit this situation in the last chapters of this book. During the intervening years, people will continue to be afflicted with the chronic diseases discussed here. Ironically, family health history offers the most cost-effective strategy not only for managing patient health, but also for performing the bioinformatics that will be necessary for employing genomic medicine in the future.

Each of us also shares about half of our own genetic variants with each first degree relative, and some of these are medically relevant. Importantly, even if a specific genetic variant cannot be pinpointed in connection with a chronic disease or condition that family members tend to develop, there is a high probability that it is shared by some of them and results in a positive family health history. Just as importantly, a person does not share about half of her/his genome with any given first degree relative, and therefore, one family member might be afflicted by the effects of a variant that is not carried by another. Further, a variant which increases adult onset disease risk may be lost solely by chance in a small family, even if the variant has no impact on reproductive success, and thereby disappears from a lineage of direct descendants. This partial sharing of variants among family members explains why diseases tend to "run in families" even if an exact genetic variant is neither known nor readily traceable. It also accounts for why some family members display symptoms or early indications of a familial condition while others do not. In practice, the value of information about apparently unaffected family members is frequently overlooked and this will be a topic for later discussion.

As one moves outward from first degree relatives or back through ancestral generations, the proportion of genetic similarity between an individual and her/his blood relatives is reduced by half with each intervening birth. Therefore, each person shares about 1/4 of her/his genome with each of their second degree relatives: grandparents, blood-related aunts and uncles, and half-siblings. About 1/8 of the genome is shared with third degree relatives—each cousin, great uncle, great aunt, and great-grandparent. As a matter of probability, of course, each person is most likely to share specific genetic variants with first and second degree relatives. Nevertheless, it is possible that someone shares a medically relevant gene variant with a distant relative (for example, one's third cousin with whom the overall genetic identity is 1/32), and therefore, both could develop a common trait, disease, or symptom even if they never knew about each other's existence. When this occurs (and it probably is more common than realized since health information from cousins and other distant relatives is infrequently obtained), a variant's threaded passage through generations may be traceable back to a common ancestor, known as a *founder*, where the change (that is, the original *mutation*) in the DNA sequence first occurred and was passed along to a child. Currently some direct-to-consumer (DTC) genetic testing services report the percent of Neanderthalic variants (those that have persisted in modern humans since the time of inter-matings between *Homo sapiens* and *Homo neanderthalis*) as part of their genetic offerings to consumers. Neanderthal variants in modern genomes clearly demonstrates the capacity for individual founder variants to persist and proliferate through hundreds of generations of descendants, especially if the variant is associated with a trait or condition that improves survival in certain environments.

Ancestry tests use the same concept, founder variants, to trace DNA variants back to ancient populations as they migrated across the globe. In most cases, these genome variants are nothing more than "typos" in the six billion base pairs and exert no functional consequence. Nevertheless, some of these typos have been catalogued and employed by commercial DNA testing services to impute familial lineages going back hundreds and thousands of years. For example, if a mutation occurred in a child born in a Norse village 1500 years ago and it was passed on by chance through the child's descendants, it could be identified and defined as a Scandinavian/northern European founder variant simply because it proliferated within a population residing in that region of the World at the time (even if it had not health impact whatsoever). It's worth emphasizing that such variants typically have nothing to do with Viking culture, Swedish meatballs, or a propensity to build longboats. In other words, variants like these do not "cause" people to become Vikings, Germans, East Africans, or Central Asians, they are simply minor changes in the genome's sequence that as already noted, first arose in a region where the progenitors of a present-day population resided. It is also worth noting that those Norse settlers already carried variants which themselves might be traceable to founders among even earlier human populations. In this respect, the occurrence of founder variants are similar to the appearance of printing errors in early edition Bibles. In the earliest days of printing, typesetting was a laborious task and inadvertent errors could appear anywhere in the text. Often, the changes did not change the meaning of the text (though in a few instances they did). Once an error occurred, it was often propagated in future printings, so that nowadays, the type and location of printing errors are used to check the authenticity of Bibles suspected or claimed to be early and historically important copies.

A few medically relevant variants actually arose in this way from these ancestral populations, such as the Factor V Leiden variant.[13] The test for this variant is among the most widely ordered genetic tests in the United States and is used to identify the genetic source of thrombosis in patients who have a family health history of deep vein thrombosis, or who have unexpected or repeated thromboses (the utility and limitations of this and other genetic tests will be discussed later). This variant apparently arose from a single founder living in northern Europe more than 20,000 years ago when the Earth's human population was less than a million. A possible explanation for its persistence is that the mild hypercoagulability caused by the variant's effect on the blood clotting protein, Factor V, improved survival during prehistoric childbirths. In the course of human medical history another founder of the same variant may have existed, but the most likely explanation for the widespread occurrence of the Factor V Leiden variant today, especially in European populations, is that current living carriers are distant descendants of that original founder.

In fact, every human birth in our history is a founder event for dozens of novel genetic variants.[14] Put another way, each one of us today carries (1) ancient variants going back to the time before *Homo neanderthalis* and *Homo sapiens* interbred, (2) variants that arose in our more immediate ancestors, (3) variants for which our own parents were the founders (and which are different than the founder variants carried by our siblings, and even identical twins), and (4) variants that are novel in us, for which we are the founders (Chapters 3 and 4). To be sure, every parent will continue this genetic tradition with each of their children. It follows that if a person inherits a haplo-genome from each parent, she/he will inherit about half of the many millions of genetic variants accumulated through one's ancestry and residing in their mother's genome, and about half of the genetic variants similarly carried in their father's genome. Genetically speaking, our ancestral history is a matter of record.

Inherited alterations of gene sequence or activity can predispose adult disease onset

Health status can be affected by a gene's sequence and/or its activity. Gene activity has been repeatedly implicated across a broad spectrum of human diseases, but it is important to understand and appreciate the difference between an alteration in gene activity (often called gene expression) and an alteration that changes the RNA sequence and sometimes, the amino acid sequence encoded by the gene. As the earlier discussion noted, changes in the DNA sequence of a gene may alter the nucleotide sequence or the amount of the specified RNA (i.e., the gene's activity). That encompasses the extent of what a gene directly causes and tempers the notion that genetic variants "cause" adult onset diseases. Certainly, a gene variant that results in a catastrophic structural change in the protein it encodes can directly lead to a disease state. For those whose adult health status is good, the relationship between genetic variants and chronic disease is circuitous and typically depends on triggering events. In other words, genetic variants usually do not "program" the onset of adult disease, but they can leave the carrier unusually vulnerable to a specific disease or condition, especially if an environmental exposure, a lifestyle choice, a second cellular event, or an undiagnosed symptom accompanies the inherent susceptibility.

The explosive increase in the use of genomic tests for a variety of medically relevant conditions fosters the impression that genetic tests are a precise form of patient risk assessment, whereas family health history is a less accurate diagnostic tool. *In reality, the use of family health history is better viewed as a comprehensive indicator of a patient's disease risk, which can sometimes be clarified further by genomic and/or genetic testing.* Blood relatives share some of their genetic variants with each other and they often share exposures to environmental factors that could impact health. A genetic test result, when used judiciously, can guide an individual's health management by defining precisely their disease susceptibility, but the relationship between genetic and environmental factors is more nuanced and complex than generally understood.

Familial disease risk is a product of both heredity and shared environment

There is sometimes resistance to the notion that inherited vulnerabilities exert an outsized influence on adult health status. In practice, it is counterproductive to propose that individuals make healthy choices, while implying that inherited susceptibilities dictate their medical fate. Unhealthy lifestyle choices and hazardous exposures can completely overshadow inherited contributions, and making healthy lifestyle choices is as important as genetic variants, even without knowing an individual's disease risk. However, knowing that risk comes with substantial benefits. For example, in some instances, a healthy choice for one person may be unhealthy or provide minimal benefit for another. For the foreseeable future, altering one's environment and behaviors, which includes obtaining screenings and preventive treatments, offers the most cost-effective strategy for reducing an inherited disease vulnerability. A major subnarrative of this book is that knowledge of one's family health history actually heightens the perceived importance of managing environmental exposures and also, of understanding environmental risks more precisely.

The effect of a given gene variant on molecular and cellular function is one that may proceed for decades before any sign of a vulnerability or impending breakdown is evident. In this respect, the effect of the variant is roughly analogous to a subtle alteration in the blueprint design and/or structure of a house, which has no discernible effect on its appearance or function, but leaves it vulnerable to the effects of water seepage. If the placement of this vulnerable structure (its environment) allows for water seepage that erodes the foundation, the problem could eventually reach a "tipping point" when serious and destructive structural damage occurs. If the condition is detected before this point, a remedy might be made that is both inexpensive and effective. Of course, if the seepage and erosion is substantial, no structure might withstand the damage caused.

On the other hand, even with a vulnerability to serious damage, it might never be noticed if the triggering environmental conditions are nonexistent or relatively minor. A substantial environmental risk factor may lead to a disease even in a person with no unusual genetic susceptibility, but it is simply logical that if one is genetically vulnerable, then the likelihood that such a person will be affected by a modest environmental risk is higher than for one who is less susceptible.

Smoking is a classic example of an important environmental exposure (lifestyles and behaviors are important components of environmental exposures) that undeniably increases the risk of several diseases, including lung cancer, and has a well-defined quantitative relationship between the dose (frequency and duration of smoking) and disease risk. The mechanistic connection between smoking and lung cancer has been conclusively established because the active ingredient in cigarette smoke causes a form of DNA damage that is distinctive and highly prevalent in lung biopsies taken from lung cancer patients who are smokers. This type of DNA damage is rare in nonsmokers.[15] With that being said, some smokers never develop lung cancer, and some nonsmokers do. So, while the environmental risk posed by smoking is unmistakable, lung cancer occurrence also apparently involves differences in susceptibility. This is sometimes ignored when assessing causes of chronic disease; for instance, a recent medical report estimated that almost half of all cancers in America today are attributable to lifestyle factors: smoking, obesity, and alcohol use. However, the obvious counterpoint is: why doesn't everyone displaying poor lifestyle choices suffer from cancer? The answer can be partly attributed to differences in inherited vulnerabilities that are often shared with other family members. This in no way reduces the importance of environment and lifestyle. The point to be made is that unhealthy lifestyle habits may raise the risk of cancer in some families. In other families, these poor choices may substantially raise the risk of heart disease, or stroke, or type 2 diabetes. By corollary, what is most urgent to do for members of a family that is susceptible to a myocardial infarction might be different than for those in a family that is vulnerable to colon cancer.

The question remains: what proportion of our susceptibility to adult onset disease is attributable to our heredity? Twin studies offer a rigorous proxy for evaluating a heritable contribution to disease risk because fraternal twins are genetically equivalent to full siblings (each of whom shares half of their parents' genetic variability) and identical twins are essentially the same genetically. Moreover, twins are also likely to show the effects of shared environment, both prenatally and postnatally. An important Scandinavian longitudinal study analyzed the impact of heredity and shared environment on the incidence of specific cancers in over 100,000 same gender fraternal and identical twin pairs over several decades.[16] Generally, if one twin developed a cancer, the probability that the other twin would develop it was significantly higher than what is seen between randomly chosen individuals. As would be expected when considering the effect of heritable factors and shared environment on cancer risk, the level of cancer concordance was even higher for identical twins, who share an entire genome, than it was for fraternal twins, who share about half. Nevertheless, heritable factors alone cannot account for the level of twin concordance for cancer. Environment also played a significant role, and its impact was inferred by the average difference in age of onset, about 8.5 years (4.5–15 years), between twins who developed the same type of cancer. In most cases, when one identical twin developed a specific cancer type, the other did not. This further argues that inherited genetic variation, to the extent that it underlies observed concordance, is *predisposing* rather than causal—a theme that will be explored further in later chapters.

The relative contribution of heritable factors and shared environment were further estimated statistically for their respective contribution to identical and fraternal twin concordance (Table 2). For prostate cancer, heritability was a significant contributor to twin concordance while shared environment exerted no measurable impact. In most other cases, shared environment was a significant source of the concordance seen between twins. For breast and colorectal cancer incidence, twin concordance was attributable to both shared heredity and environment. Not surprisingly, shared environment was the predominant contributor to twin concordance for lung cancer incidence. A substantial portion of twins were scored as discordant in this study because they developed different types of cancer. This discordance can be partly attributed to the recognition that a given gene variant carried within a family sometimes leads to cancers in different cell types, and an understanding of the relationship between genetic variation and the range of possible cancers attributable to a given **hereditary cancer syndrome** is an important consideration when analyzing a family's medical history that will be explored later. From a practice standpoint, the co-incidence of cancer seen in twins further strengthens the rationale for assessing disease risk from the medical histories of one's first- and second-degree relatives. In other words, the vulnerability to specific cancer types is often shared among family members and by tabulating these occurrences, a person's relative vulnerability to a specific type of cancer can be estimated—with the caveat that some predisposing variants can underlie different cancer types within a family. As will also be seen, not every carrier of even highly pathogenic variants will actually develop cancer.

The generalizability of the twin study's cancer incidence has limitations when contemplating its importance for clinical practice broadly. First, the subjects were taken from a relatively homogeneous genetic population occupying a limited geographic range and culture. It follows that the study analyzed a relatively small fraction of the total genetic variation that exists in the human population with little diversity in cultural, ethnic, biophysical, or environmental contributions that might modulate cancer risk. If a similarly designed twin study was carried out in a different and relatively isolated population (that is, one with limited migration entry from genetically unrelated populations), then cancer incidence and the effects of heritability and shared environment might also be

TABLE 2 Estimates of heritability and shared environment contribution to familial risk for incidence of specific types of cancer occurring in monozygotic and dizygotic twin pairs (Nordic Twin Study of Cancer (*NorTwinCan*)).

	Heritability % (95% CI)	Shared environment
Overall cancer	33 (30–37)	–
Head and neck	9 (0–60)	26 (0–65)
Stomach	22 (0–55)	6 (0–31)
Colon	15 (0–45)	16 (0–38)
Rectum and anus	14 (0–50)	10 (0–38)
Lung	18 (0–42)	24 (7–40)

Skin
  Melanoma
  Nonmelanoma
	
	58 (43–73)
	43 (26–59)
		
		0
		0

Breast	31 (11–51)	16 (0–31)
Corpus uteri	27 (11–43)	0
Ovary	39 (23–55)	0
Prostate	57 (51–63)	0
Testis	37 (0–93)	24 (0–70)
Kidney	38 (21–55)	0
Bladder, other urinary organs	30 (0–67)	0
Leukemia, other	57 (0–100)	0

different. This observation itself is highly susceptible to misinterpretation about general features of ethnicity and race as well as their effects on individual disease risk that will require further exploration. Secondly, the elements which comprise "shared environment" were not defined for the twin study. The study design simply noted that the composite environment was shared between twins and might be interpreted to suggest that a specific environmental factor is at play without assessing it directly. As genetic information is more precisely analyzed within a family, a concomitant rigor will need to be applied to an analysis of potential environmental triggers if the goal is more precise medical practice. If beef consumption raises the risk of colorectal cancer, what aspect of beef consumption is primarily responsible? Is it how often beef is consumed? How it's prepared? Did the cattle's diet (e.g., pesticide consumption) affect human cancer risk? Is the increased risk from beef consumption great enough per se to make it a central feature of one's cancer prevention strategy? If the family's history suggests a significant vulnerability to type 2 diabetes and no history of cancer, what disease prevention strategy is most effective? Assuming that even the most conscientious among us cannot eliminate all our risks, is limiting beef consumption worth the effort for an individual whose familial risk is type 2 diabetes? These questions raise a more general concern about how the disease risk impact of environmental factors are defined, as my physician highlighted when he questioned my explanation for my stroke with: "What's stress?"

Whatever health effects an environmental factor may cause on its own are initiated through molecular changes in cells, that in turn, affect the regulation (and dysregulation) of genes, because gene activity is as intrinsic to a cell's function as having electricity is to living in a modern home. Many common tests that measure a variety of biomolecules associated with disease status, including inflammatory markers, high and low density lipoprotein cholesterol, prostate serum antigen, triglycerides, and blood glucose levels, also assess gene activity and gene product (i.e., protein) function. In this respect, family health history sometimes offers important clues about what lies beneath an unusual or concerning test result and even could reveal a pattern of clinical test results that is familial.

It remains true that the effects of environment on a person's health also depends upon their own body's response to the challenge. We have already noted that one or more genetic variants shared by some family members might affect the response to a given environmental challenge, as evidenced by an effect on a gene's activity or the modulated function of its protein product. However, a host of **epigenetic mechanisms** that affect a cell's biology without altering DNA sequences have been identified over the last few decades and these mechanisms, triggered by one or more environmental factors, may eventually prove to be "familial" when environmental triggers are shared among family members. Note that epigenetic simply means changes to DNA that do not affect the nucleotide base pairs. In these events, the DNA sequence is unchanged, a "T" is still a "T," but other molecules and proteins may attach to it causing changes in function. The extent by which epigenetic factors contribute to overall human health, and contribute to adult disease onset is a subject of growing interest though well-described molecular mechanisms tied to adult health status have not yet been fully elucidated. The possible connection between epigenetics and disease, including various forms of cancer, is a subject of active research and is discussed further in Chapter 9.

At least two important inferences about epigenetic contributions are relevant when evaluating the utility of family health history. First, family health history reflects both epigenetic and genetic influences on a patient's disease risk. Secondly, epigenetic processes triggered by environmental factors *and* genetically based vulnerabilities can coexist to elevate a specific disease risk, though the relative contributions remain to be sorted out for specific conditions and families.

In other words, epigenetic factors do not exclude the simultaneous impact of genetic factors on a patient's risk for a specific disease, and vice versa.

Collection and analysis of useful family health history poses challenges for healthcare

The relationship between specific genetic variants and serious diseases including sickle cell anemia, cystic fibrosis, Huntington's disease, and specific forms of cancer justified the effort to characterize the entire DNA sequence of the human genome, through the Human Genome Project, which was completed ahead of schedule and under budget, owing to the effort and commitment of thousands of researchers around the world. The product of this successful effort is now accessible to everyone through a publicly available database (www.ncbi.nlm.nih.gov). It established that all humans share a high degree of similarity with the consensus reference human genome sequence, but that everyone also carries a considerable number of DNA variants, some of which could influence disease risk. The practice of genomic medicine, however, continues to lag in everyday healthcare, perhaps because a gap between the science and the application continues to exist.

When introducing a genomic medicine initiative several years ago, from which many insights were drawn upon for this book, several hospital administrators, support staff, and patients were asked what the term "personalized medicine" meant to them. Our assessment team was surprised to see that many of the responses referred to friendly and responsive personnel, clean and attractive facilities, amenities, and on-time appointment schedules. Those responses convey the expectations of the healthcare consumer, but say nothing about the need for an active engagement and exchange of information between provider and patient that is at the core of the vision for precision medicine.

The survey responses highlight the need for an essential ingredient for the true success of a medical practice intended to keep patients healthy, namely, they must be guided through a process geared toward understanding their disease vulnerabilities and then encouraged to take the appropriate and *effective* actions to maintain their own health (while respecting the possibility that a patient's choices may involve more than a quantitative risk assessment). The medical benefit of gathering this information ultimately depends on finding interventions and treatments that keep an individual healthy and minimize the costs of reactive treatment, and family health history can provide a roadmap for preventive intervention.

Unfortunately, as we discovered when promoting the collection of family health history in primary care, physicians are uncertain about what family health history information to collect, how to collect and analyze it, and how to act on it once it is analyzed. When we pored through medical records in a primary care practice that agreed to provide information, we found that only 4% of almost 400 patient records contained a sufficient level of family health history information just to carry out a patient's disease risk assessment.[17] Moreover, none had included all the updated information ideally compiled to analyze disease risk. When I asked one physician what happened to the intake forms that typically ask the patient to report whether they or someone in their family ever suffered from one of several listed conditions, he responded that the completed form is typically tucked into the medical record. Then, he asked me rhetorically: "Do you know why we collect that information?" He went on to show me the Centers for Medicare & Medicaid Service's provisions for medical reimbursement, which includes the requirement that the provider "collect the family history" at the time of initial patient intake.[10] It goes without saying that if the overriding purpose of collecting a family health history is to comply with reimbursement requirements, there's little incentive to do more, especially in the absence of vetted guidelines for its collection.

This lack of knowledge about family health history could lead to some counterproductive and avoidable consequences. During my own rehabilitation, I was examined once by a resident who recommended I take aspirin. I passed along my acquired wisdom by gently reminding him that I had not experienced "*that* kind of stroke." I became more irritated by my neurologist's response when I reported my family's history of hemorrhagic strokes. He noted that the only patients he had seen with hemorrhagic strokes were cocaine addicts, as he cast an inquisitional glance at me. I sarcastically responded that I couldn't remember the last time I had taken cocaine, and I was even more upset about it than I conveyed, because I have known lives ruined and lost by drug and alcohol abuse. After my appointment, however, I contemplated the neurologist's reaction further. As I processed his comments, it became plausible to me that his only encounter with hemorrhagic strokes, especially in physically active patients with moderate BMIs, might include substance abuse. Conversely, is it further possible that only a small fraction of substance abusers actually suffer a subarachnoid hemorrhagic stroke, that is, only those who have an inherent vulnerability, as evidenced by their own family history? It must be added that my neurologist had no access to information about my family's history to surmise that my condition truly is a "family matter," and that my own consciousness of the connection was vague until I concentrated my efforts. My case also illustrates the difficulty with recognizing the importance of family history until *after* a health event or disease onset in an individual, which arguably illustrates the need to recognize early warning signs and obtain precise medical information, especially when other relatives were afflicted at a comparatively early age.

The deficits in awareness and documentation of family health history will not simply abate with time or with the transition to electronic medical records. In fact, as genomic testing and highly sensitive molecular tests continue to develop, the problem might actually worsen amid a burgeoning flow of test results lacking the accompanying clinical information needed to properly interpret the information. We find ourselves dealing once again with fundamental questions: Is family health history information obtained from an individual robust enough to guide a personalized intervention program? Can it be compiled and used without disrupting clinical patient flow? To make this work, tools and clinical operations that allow for rapid, accurate, and updatable family health history records will be critical, and this topic will be addressed in later chapters. As will be noted, the capability to analyze genomic and other molecular test results will increasingly depend upon having health information about the patient and his/her genetic relatives.

The utility of family health history will depend upon new tools and updated models of clinical patient flow

The dilemma posed for me and my relatives also highlights a more general obstacle, namely, the need for education and engagement when making decisions based on family medical history and genetic/genomic test results. As will be noted, the need for accuracy and clarity is important, as well as awareness of the issues that surround both the processing of family health history and genetic information, and ultimately, taking deliberate and consented actions based on the choices that follow. The genetic counseling profession arose in the 1970s to fill the gap between what medical results say and what patients do in response. Genetic counseling was originally focused on patients dealing with the diagnosis of, or risk for, prenatal and perinatal genetic disorders. As genetic testing expanded into adult diseases, the need for genetic counseling geared toward adults concomitantly increased. When considering the utility of family health history, the entire patient flow in primary care depends upon an ongoing relationship between the practice and the patient, from the time of intake, through examination, and in the follow up, which includes genetic counseling sessions, when appropriate, specific recommendations, and ongoing updates of family health. Without an effort to sustain this relationship, much of the value of family health history is not realized. Part of this book will be devoted primarily to a discussion of implementing the use of family health history in primary care practice.

The advent of whole genome sequencing and refined diagnostic tests have significantly improved the precision of clinical diagnostics in recent years, and the pace of clinical implementation of these tools will likely advance and quicken in the coming years. The last chapters of this book will focus on these trends and some of the implications of these tools for improving health outcomes and controlling the cost of patient management. Nevertheless, careful analysis of family health history can increasingly serve as a critical source of insight about a patient's health status and health prognosis.

Summary and conclusions

Family health history has long been recognized as a source of useful information when assessing a patient's risk for developing adult onset diseases or experiencing debilitating health events. Family health history is an important and underutilized source of information to guide patient management as more information is gathered about the molecular and cell biology of disease onset, diagnostic tests are employed for monitoring patient health status, and genetic variants that are predisposing for individual disease risk are identified. The summarizing conclusions drawn from this chapter are:

- Family health history provides highly relevant and useful information to the provider, the patient, and her/his genetically related relatives concerning personal risk for experiencing adult onset diseases and/or medical events that negatively affect health status.
- Common adult onset diseases exert a substantial effect on lifespan and quality of life in the population.
- FHH information is useful for identifying recommendations that prevent or delay disease onset via cost-effective interventions and treatments.
- Every human carries millions of DNA sequence variants that include ancient variants, variants that arose more recently in family lineages, and novel variants associated with one's own conception. A few of these potentially increase the risk of adult onset disease.
- Genetic variants rarely lead directly to adult onset disease, but instead, are generally predisposing and thereby increase the risk for an adverse change in health status influenced by specific lifestyle choices and environmental exposures.
- Symptoms or indications often arise before a familial health condition or disease appears, thus allowing for an intervention strategy to be employed.
- Family health history can sometimes be associated with specific genetic variants, but usually genetic test results are not necessary to formulate an intervention or treatment regimen to reduce a patient's personal disease risk.
- Collection and analysis of family health history in primary care is impeded by a variety of obstacles including a lack of patient knowledge about family health history, provider uncertainty about the utility of family health history for follow up diagnosis, monitoring, and patient management, and a lack of collection and analytical tools that do not impede patient flow.
- The value of family health history in medical care will depend upon the acquisition and analysis of more and accurate family health information, providers who are prepared to interpret and act on family health history, improved patient education, and in practice, an emphasis on maintaining a patient's good health status.

Vincent C. Henrich is the lead author for this chapter. First person statements are from his point of view.

References

1. Leading Causes of Death. https://www.tomwademd.net/wp-content/uploads/2018/04/2015-Chart-On-Causes-Of-Death-From-CDC.png; 2015.
2. National Center for Health Statistics. *Health, United States 2016: With Chartbook on Long-Term Trends in Health*. Hyattsville, MD: National Center for Health Statistics; 2017.

3.   National Center for Health Statistics. *Health, United States.* https://www.cdc.gov/nchs/data/hus/2017/015.pdf; 2017.

4.   Murphy SL, Xu JQ, Kochanek KD, Arias E. *Mortality in the United States, 2017.* NCHS Data Brief, no 328, Hyattsville, MD: National Center for Health Statistics; 2018.

5.   Ruby JG, Wright KM, Rand KA, et al. Estimates of the heritability of human longevity are substantially inflated due to assortative mating. *Genetics.* 2018;210:1109–1124. https://doi.org/10.1534/genetics.118.301613.

6.   Sebastiani P, Andersen SL, McIntosh AI, et al. Familial risk for exceptional longevity. *N Am Actuar J.* 2016;20:57–64.

7.   Christensen K, McGue M, Petersen I, Jeune B, Vaupel JW. Exceptional longevity does not result in excessive levels of disability. *Proc Natl Acad Sci USA.* 2008;105:13274–13279.

8.   Bor AS, Rinkel GJ, van Norden J, Wermer MJ. Long-term, serial screening for intracranial aneurysms in individuals with a family history of aneurysmal subarachnoid haemorrhage: a cohort study. *Lancet Neurol.* 2014;13:385–392.

9.   Geer EB. *The Hypothalamic-Pituitary-Adrenal Axis in Health and Disease: Cushing's Syndrome and Beyond.* Switzerland: Springer International Publishing; 2017, ISBN: 978-3-319-45950-9. 327 pp.

10.   Boughey JC, Hartmann LC, Anderson SS, et al. Evaluation of the Tyrer-Cuzick (International Breast Cancer Intervention Study) model for breast cancer risk prediction in women with atypical hyperplasia. *J Clin Oncol.* 2010;28:3591–3596.

11.   Wang X, Huang Y, Li L, Dai H, Song F, Chen K. Assessment of performance of the Gail model for predicting breast cancer risk: a systematic review and meta-analysis with trial sequential analysis. *Breast Cancer Res.* 2018;20:18. https://doi.org/10.1186/s13058-018-0947-5.

12.   Brentnall AR, Harkness EF, Astley SM, et al. Mammographic density adds accuracy to both the Tyrer-Cuzick and Gail breast cancer risk models in a prospective UK screening cohort. *Breast Cancer Res.* 2015;17(1):147. https://doi.org/10.1186/s13058-015-0653-5.

13.   Kujovich JL. Factor V Leiden thrombophilia. *Genet Med.* 2011;13:1–16. https://doi.org/10.1097/GIM.0b013e3181faa0f2.

14.   Palamara PF, Francioli LC, Wilton PR, et al. Leveraging distant relatedness to quantify human mutation and gene-conversion rates. *Am J Hum Genet.* 2015;97:775–789. https://doi.org/10.1016/j.ajhg.2015.10.006.

15.   Alexandrov LB, Ju YS, Haase K, et al. Mutational signatures associated with tobacco smoking in human cancer. *Science.* 2016;354:618–622.

16.   Mucci LA, Hjelmborg JB, Harris JR, et al. Familial risk and heritability of cancer among twins in Nordic countries. *JAMA.* 2016;315:68–76. https://doi.org/10.1001/jama.2015.17703.

17.   Powell KP, Christianson CA, Hahn SE, et al. Collection of family health history for chronic diseases in primary care. *NC Med J.* 2013;74:279–286.

18.   Evaluation and Management Services. *CMMS, DHHS.* 90pp, https://www.cms.gov/Outreach-and-Education/Medicare-Learning-Network-MLN/MLNProducts/Downloads/eval-mgmt-serv-guide-ICN006764.pdf; 2017.

Further reading

Hood L, Flores M. A personal view on systems medicine and the emergence of proactive P4 medicine: predictive, preventive, personalized and participatory. *N Biotechnol.* 2012;29:613–624. https://doi.org/10.1016/j.nbt.2012.03.004.

Hood L, Friend SH. Predictive, personalized, preventive, participatory (P4) cancer medicine. *Nat Rev Clin Oncol.* 2011;8:184–187. https://doi.org/10.1038/nrclinonc.2010.227.

Chapter 2

Family health history's place in genomic medicine

Lori A. Orlando, Brian H. Shirts, and Vincent C. Henrich

- What is family health history?
- Family health history as a risk marker.
- The value of family health history for informing clinical care.
- What are the levels of risk defined by family health history and what do they mean?
- Clinical validity and utility of family health history.
- Family health history's relationship to genetic testing.

The term genomic medicine came in to use around 2011, initially as a synonym for "precision medicine", and was meant to encompass all things genomic. But as our understanding has evolved, so has the definition. Today the National Human Genome Research Institute defines genomic medicine as "an emerging medical discipline that involves using genomic information about an individual as part of their clinical care (e.g., for diagnostic or therapeutic decision-making) and the health outcomes and policy implications of that clinical use" (https://www.genome.gov/27552451/what-is-genomic-medicine/). Thus, the term is now reserved for translation into clinical environments and not for bench or discovery research. But I should probably take a step back and ask, what does "genomics" mean? If you are like me, the concept brings to mind amorphous images of double helices, but not a clear picture of how it differs from "genetics." From the vantage point of how the two relate to family health history, genetics refers to a specific gene/gene variant and genomics to all the genes and how they interact (Fig. 1). For example, if my patient has a strong family health history of breast cancer that warrants *BRCA1* and *BRCA2* gene evaluation for Hereditary Breast and Ovarian Cancer syndrome, I am considering a genetic test (i.e., a test of a single gene). In fact though, there are a very large number of genes that are now known to contribute to familial and hereditary forms of breast cancer: *BRCA1*, *BRCA2*, *PALB2*, etc. and, therefore, it is more likely that I will perform a genomic test of somewhere between 5 and 45 genes depending upon which test I select. In the next chapter we will go into more detail about the

Managing Health in the Genomic Era. https://doi.org/10.1016/B978-0-12-816015-2.00002-X

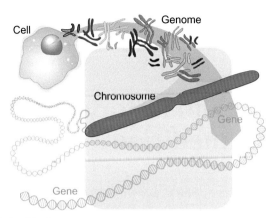

FIG. 1 The relationship of genes to genomes.

relationship between family health history and genetics. At the time of writing (2019) genomics is still a rapidly evolving field, prohibiting us from providing clear guidance on what tests to perform for a given family history; however, there is good news, genetic counselors fill this gap by serving as a bridge between the state of the science in genomics research and clinical care. They are our friends and if we do not already have a relationship with one, we should find one and keep them close!

My close relationships with genetic counselors started in 2008. At that time, I was practicing internal medicine and had completed a health services research fellowship. I had learned about family health history data collection in medical school and thought I was doing a good job of collecting and understanding it. When I was in medical school, studies about treatment options for those with pathogenic *BRCA1* and *BRCA2* variants were just being published and we appreciated that family history of breast or colon cancer onset at a young age was "bad," as was early onset heart disease. Otherwise we didn't know much. For the most part we presumed, as we had been told, that hereditary conditions were uncommon and that our patients' family health history information would be unrevealing. As a conscientious doctor, I religiously asked my patients about their family health history, focused on those three factors and on occasion I suggested an early colonoscopy or better cholesterol control, but never once had I referred a patient to a genetic counselor or considered that perhaps some of the patients I so diligently cared for might need genetic testing of some type. Fast forward to 2008 when I was asked to lead a study in family health history driven risk assessment in primary care practices (the Genomedical Connection Study), where I met several amazing genetic counselors and my extraordinary co-author, Vince Henrich. This study changed everything. I became an unapologetic family health history crusader and have never looked back!

During the Genomedical Connection Study, my team assimilated a great deal of information about the relationships and nuances of family history. For example, multiple endocrine neoplasia, is a syndrome I learned about in medical school. However, there the context was—if your patient develops thyroid cancer, ask about these other diseases that could suggest this syndrome—which by the way is incredibly rare. It never occurred to me that we should routinely ask about a family history of these different conditions, given they could suggest a predisposition to multiple endocrine neoplasia in the family, or that genetic testing might help us take better care of our healthy patients. Working with my colleagues, we explored the idea that technology could facilitate more accurate collection of family health history from patients, analyze the information, and provide feedback to clinicians and patients about the level of risk across a spectrum of diseases and tie it to evidence-based recommendations for how to manage that risk. The goal was to optimize a family health history based risk assessment across the general population and then tailor their preventive care to their risk level.[1] In a later chapter, I will talk more about methods for streamlining this process in clinical practice; but in this study the key findings for me were: (1) about 20% of healthy patients in clinical practice met criteria for genetic counseling[2]; and (2) the providers, like me, all felt they were doing a great job assessing family health history before the study; but realized during it that there was significant room for improvement. We were all amazed at how many patients were being identified and the vast majority were gliding along, under our radar, undetected. In fact, the findings were so dramatic that the providers and the health system decided to continue the study, at their own cost, one year after funding ended. This one study changed my entire outlook on risk assessment and family history. It also led me towards a career bridging real world clinical providers with the rapidly advancing field of genomics. It took me years to really begin to understand genomics and SNPs and variants of undetermined significance (more about these later). What I am here to tell you is that you do NOT need to know these things to take good care of your patients. Focus on identifying those who are at risk using family health history based risk assessments and have your genetic counselor colleague handy to guide you through the next steps.

What is family health history?

Okay with that introduction to genomics and genomic medicine, let's consider what family health history is and how it fits into genomic medicine. You may be thinking "duh!" it's the health history of an individual's relatives. And, of course, you would be correct, but what makes up a good family health history? How many relatives, what type of relatives, what information do you need to know about each? If you already know the answers to these questions, then skip to the next paragraph! If you don't, then keep reading. To start, a brief review from Chapter 1 (Fig. 2). **First degree relatives** are those that you are most closely related to and with whom you share 50% of your DNA. This includes your parents, full-siblings, and children. **Second degree relatives** are

Collection
Degrees of Relationship

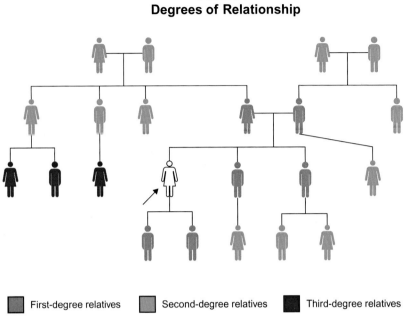

First-degree relatives   Second-degree relatives   Third-degree relatives

FIG. 2 The relatedness of relatives and their relationships to each other.

those with whom you share 25% of your DNA. They are the first degree relatives of your first degree relatives, which includes your grandparents, aunts and uncles, nieces and nephews, grandchildren, and half-siblings. **Third degree relatives** are those with which you share 12.5% of your DNA and include great-grandparents, great-grandchildren, and first cousins.

With this in mind a high-quality family health history should include[3]:

(1) first, second, and third-degree relatives.

(2) each relative's lineage (the side of the family they come from—maternal or paternal).

(3) relatives' gender or, if transgender, birth gender to current gender (e.g., male to female).

(4) collect all conditions affecting all relatives or if medical information is incomplete specifically indicate conditions not occurring in the family (e.g., negative family history of breast cancer).

(5) age of onset for all reported conditions.

(6) for living relatives indicate their age.

(7) for deceased relatives indicate their age of death and cause of death.

Lastly, the family health history should be kept up to date. In an interesting 2011 study in the Journal of the American Medical Association, researchers enrolled participants from a cancer registry to assess how often their family history changed and whether it impacted their cancer risk. They concluded that family health histories should be updated at least every 5 years.[4] While this assessment was focused upon cancer, it seems reasonable to assume that relevant changes in family history for other conditions, such as diabetes and coronary artery disease are likely to occur at least as often. With this information in hand you can now start to collect high quality family health histories!

But alas, asking for all this information from patients can be fraught with misunderstandings and confusion. Most of us do not know much about our family's health history. Even physicians rarely can provide this much family history data without preparing. If we can't do it, we certainly should not expect our patients to! Just like we may need to consult with our family members or the family historian (that one person who knows everything about everyone in the family) to learn about medical conditions affecting our relatives, our patients could benefit from a heads up about the type of information needed and some guidance on how to talk with family members about it. In addition, there are a few common pitfalls that we should try to address before sending them out to gather their data.

- Biology—ensure they collect data on only biological relatives. You would be amazed at how many times people are unaware that Aunt Suzie is just a close family friend or a second cousin once removed and not really an "Aunt." In a study conducted by the authors, 100 patients providing a family history during an office visit were given education about collecting family history and invited back 2 weeks later to update it. Among the 100, 24 added or <u>deleted</u> at least one relative and the mean # of relatives changed per person was 4[5]!

- Cancer—ensure that when a cancer type is reported that it represents only the site of origin and not the site of metastases. For example, it is not unusual to hear that uncle Bob had bone cancer; but primary bone cancer is extremely rare. The majority of the time, bone cancer is actually prostate or breast or some other primary tumor that has spread to the bones.

- "Female" cancers—the female reproductive system is confusing. For the general population the difference between the cervix, uterus, and ovaries is not clear and/or not important. However, since we know that cervical cancer is caused by a virus acquired through sexual encounters, and both uterine and ovarian cancers are frequently associated with hereditary cancer syndromes the distinction is critical. Providing some guidance about identifying the exact organ can greatly enhance the accuracy of risk assessment.

- Heart disease—this can be very tricky because the myriad types of heart conditions generally are all interpreted by the general population as the same. So atherosclerotic coronary artery disease is reported in the same way as an arrhythmia. Asking a few questions to clarify symptoms and sometimes types of treatment the relative is on can help to distinguish these.

- Colloquial names—there are numerous colloquial names for medical conditions, such as high blood for hypertension or sugar for diabetes. Being aware of these names can facilitate communication and more accurate health history collection.

By the way, Thanksgiving is National Family History Day, a great time to encourage patients, friends, and colleagues to talk to their families about their family health history. What better opportunity than a family-oriented holiday to share information with family members.

We have put together a number of publicly available resources to assist with family health history data collection at the back of this book. Some resources are designed as tools for you, to augment the knowledge imparted in this book, and others are designed for patients, to help them in their endeavor to gather pertinent information. The patient resources are designed for you to share as needed with your patients (or for you to review if the reader is a non-clinician). The ones referenced are only the tip of the iceberg, there are many wonderful resources available, but these have been reviewed carefully for both the quality of the content and the communication style. I know, you are thinking to yourself, how on earth am I going to fit this into my busy schedule. In a later chapter, we discuss different methods and tools to facilitate data collection and even some that include clinical decision support to help with that next step—what do I do now.

The value of family health history for informing clinical care

So now we know some useful information about what to ask for and a little bit about how to help our patients give us good information—but why are we doing this (other than because we were told to in medical school)? There are some very good reasons to want to incorporate family health history into our clinical routine. First, a reminder. Family history tells us about two very important aspects of our patient's lives—their biology AND their environment/behaviors. On the biology side, we already know we share DNA with our relatives and that variants can be passed through families, increasing (or decreasing) risk for certain medical conditions. These variants are not determinant—they almost never confer a 100% risk of developing a condition. Depending upon the specific variant, the risk can range from negligible (1–2% increase in risk for type 2 diabetes mellitus for one of its associated variants) to profound (60% increase in risk for breast cancer with a pathologic BRCA1 variant).[6,7] On the environmental/behavioral side, families (often) live in a shared environment. They are exposed to the same social determinants of health (such as low-quality housing, food insecurity, lack of safety), the same environmental exposures (such as lead paint or asbestos in the house, arsenic contaminated water, second hand smoke), and the same culture (excessive TV watching, limited exercise, diet high in fried foods, and so on). We know that all these factors contribute to disease risk, so sometimes it can be hard to tease out if a high incidence of coronary artery disease in a family is due to genetics, environment, or both. It is important to remember this dual impact and to take it in to consideration when contemplating the meaning of an individual's family health history. In the following chapters we will discuss how each topic, genetics and environment, relates to family health history in more detail.

Now for the "why are we doing this" part. Family health history has some extraordinary benefits for patient care. In particular family history is the single most accurate predictor of disease risk available in routine clinical care. It therefore, sits squarely on the far-left hand side of the disease spectrum—in the still healthy and we want to stay that way part. Once we develop a disease, family history is far less useful, except in the case where it can be used to decrease the risk of developing another disease. Real world examples of this abound. In one example, a young woman in her late 30s developed uterine cancer. Her cancer was treated and she was cured. At that point she was discharged from the oncology clinic and returned to her primary care provider. This young woman had a remarkable family history of colon cancer, affecting almost every relative on her mother's side in their early 50s. In addition, her mother had developed both colon and uterine cancer later in life. So, while she had developed uterine cancer and her family history could not provide any additional insight for the treatment of that cancer, the family history was highly suggestive of Lynch syndrome. In fact, she underwent genetic testing and was found to have a pathogenic gene variant in one of the Lynch syndrome genes. The importance of the family history in understanding her additional risk was paramount, and if recognized earlier could have prevented the initial uterine cancer. Her diagnosis though, enables her to manage her risk for other cancers such as renal cell, colon, stomach, and more. These interventions will help her live a longer and healthier life than she otherwise might have.

This is just one example. I have seen many cases like this in my practice now that I am highly attuned to family health history information. And while each hereditary syndrome is generally considered to be rare, we now know that they are far more prevalent than previously suspected, and in aggregate a not insignificant proportion of our regular patients are likely to have one of these syndromes. One of the reasons that prevalence has been underestimated is that genetic testing has largely been limited to those with dramatic family histories. The costs and ethical, legal, and social concerns about genetic testing have previously confined it to very specific situations. However, recently, as the technology and our understanding of its implications have matured, there have been studies conducted in unselected "healthy" populations. These studies have identified higher than expected prevalences for pathogenic variants. For example, several studies reported the prevalence of Hereditary Breast and Ovarian Cancer syndrome to be 2.5–4%, and that is just for known variants in two genes (*BRCA1* and *BRCA2*) (https://www.geisinger.org/mycode-results). There are likely to be many more variants in those genes that we do not yet know about, and there are other genes that also contribute to hereditary risk, such as *PALB2*.[8] I will discuss more about *PALB2* a little later in this chapter. In addition, family health history studies also support a higher prevalence of hereditary disease than previously appreciated. As mentioned earlier, the Genomedical Connection study identified a large proportion of two primary care populations that met criteria for genetic counseling, and similar studies across the country, by us and others, have confirmed this result.[2,9,10] All together these studies consistently suggest that we have under appreciated the risk of hereditary syndromes in the general population.

In the areas of diagnosis, prognosis, and selecting optimal therapy family health history doesn't have much to offer; but in the realm of risk assessment it can't be beat! It also has the additional benefit of supporting discovery research. Data linking family history, genetics/genomics, and disease outcomes supports a positive feedback loop that informs clinical care, refines guidelines, and further strengthens the role of family history in clinical care. And because family health history reflects BOTH genetics and environment (and their interaction) it will not become obsolete, even in a world where every baby has their genome sequenced at birth (a world that is far far away from the time of writing this book). I'll explain.

In clinical practice family health history is the most accurate and readily accessible method for assessing risk on a wide variety of medical conditions, from colon cancer to depression. Several examples described in this chapter are related to cancer, but there are many diseases for which family health history is beneficial. Examples of conditions and how they confer risk are provided in the next section.

As previously alluded to, the level of risk associated with any given combination of family health history can vary dramatically, even within one disease, and family history of a disease can increase risk for that disease, but also, in some cases, for other diseases as well. Clear as mud, right? Let me give you an example using pancreatic cancer. A family history of pancreatic cancer can be associated with an increased risk of pancreatic cancer (Familial Pancreatic Cancer), and, in combination with other cancers, can be indicative of (1) Hereditary Non-polyposis Colorectal Cancer (Lynch Syndrome) which would increase risk for colon and several other cancers, (2) Familial Atypical Multiple Mole Syndrome increasing risk for melanoma, (3) Multiple Endocrine Neoplasia type 1 increasing risk for parathyroid and pituitary cancers, and (4) Hereditary Breast and Ovarian Cancer increasing risk for breast and ovarian cancer. There are no other risk markers available that can give you this much information about so many different conditions.

What are the levels of risk defined by family health history and what do they mean?

In the example above, I mention a variety of hereditary syndromes. Let's take a minute to review what hereditary means in terms of risk and how levels of

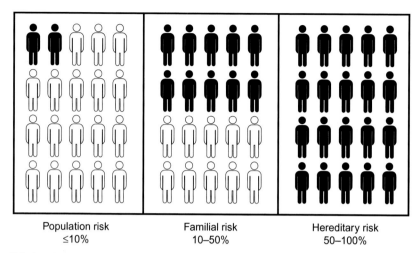

| Population risk | Familial risk | Hereditary risk |
| ≤10% | 10–50% | 50–100% |

FIG. 3 The three levels of disease risk: population, familial, and hereditary.

risk are categorized. There are essentially 3 levels: population, familial, and hereditary. In general, these go from lowest level of risk to highest, however, there are no clear-cut distinctions between where one category ends and the next begins (Fig. 3).

- Population risk level is the risk in the general population when you average together all individuals in that population. For example, the population level risk for lifetime breast cancer in the United States is 12% (1 in 9 women will develop breast cancer in their lifetime) (https://seer.cancer.gov/). However, some women in this population will have a risk level that is much greater than 12% and others will have a risk level that is much less. If an individual has a level of risk that is around 12% or lower they are considered at population level risk and undergo the usual population-based breast cancer screening with mammography.

- Familial risk level exceeds population level but does not extend as high as hereditary risk (about 12–50%). This is generally a very broad range of risk levels that is made up of everything that is not population level or hereditary level. Sticking with the breast cancer example, familial risk includes those who have a lifetime risk of breast cancer ≥ 20% (in which case they meet criteria for breast MRI screening as part of their breast cancer surveillance)[11] or a 5-year risk of breast cancer ≥ 1.66% (in which case they meet criteria for taking tamoxifen or raloxifene to prevent breast cancer)[12]; but do not have a family history suggestive of a hereditary syndrome. In many cases familial risk level is defined by risk estimates that combine family history with personal and behavioral risk factors for disease. For example, the Gail Model,[13] used to calculate 5-year risk of breast cancer, includes risk factors such as "history of chest wall radiation" and "history of breast biopsy" (along with others) in addition to a family history of breast cancer. Another example frequently encountered in primary care is when a first degree relative develops colon cancer at age 50 or less. Instead of starting colon cancer screening at 50, as occurs for those at population level risk, colonoscopies are recommended to start 10 years earlier than the relative's age of colon cancer onset.

Within the familial risk level there is a sub-category termed "common complex disease risk." This risk level does not incorporate family history at all but rather relies entirely on the results of genomic testing. Here the term "complex" means that it takes more than one gene variant to increase the risk for the disease. These complex patterns are frequently seen in the common chronic diseases, especially the ones that tend to be multi-factorial like hypertension, diabetes mellitus, schizophrenia, non-hereditary forms of prostate cancer, and coronary artery disease. Because each individual gene variant, referred to as a SNP (single nucleotide polymorphism (we will cover this topic in more detail in a later chapter)), confers only a very small increase in risk, the SNPs are combined into a "polygenic" risk score (Fig. 4). Depending upon the disease, there can be up to 100 different associated SNPs. To calculate the risk score, all disease associated SNPs are analyzed and the (small) risks associated with each SNP are summed into a single "polygenic" risk score. Even at their highest levels these polygenic scores do not reach the level of risk associated with hereditary syndromes and typically fall on the lower end of the familial risk category. However, as I mentioned earlier the field of genomic medicine is growing rapidly and our knowledge is changing on a day to day basis. There is no reason to believe that in time, as we learn more about our DNA, that they may become stronger predictors of risk.

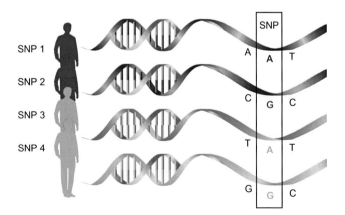

POLYGENIC RISK SCORE = SNP 1 (T) + SNP 2 (G) + SNP 3 (A) + SNP 4 (C) = 2.1

FIG. 4 Single nucleotide polymorphisms (SNPs) and polygenic risk scores.

Hereditary risk level is the highest level of risk and is associated with highly penetrant single gene variants. Like Mendel and his peas, these are generally autosomal dominant variants (only one DNA strand needs to be mutated to increase the risk of the disease), but in some cases are recessive, requiring both strands to be mutated before increasing the risk for disease. As a reminder penetrance is another way of describing risk—highly penetrant means if you have the variant you are very likely to develop the disease—so these are variants you do not want to have! For many of the hereditary syndromes, expert committees have defined family history patterns that raise concern and warrant genetic testing. Currently the only way to determine if an individual warrants genetic testing for a hereditary syndrome is via their personal and family history, though new guidelines based on the results of tumor testing in those who have cancer are emerging. In our example of breast cancer, the National Comprehensive Cancer Network has published a 60-page document outlining when to suspect Hereditary Breast and Ovarian Cancer. It includes criteria like one or more close relative with breast cancer at age 50 or less, OR two or more close relatives with breast cancer and/or pancreatic cancer OR a combination of breast cancer with any one of the following: thyroid cancer, prostate cancer, sarcoma, adrenocortical carcinoma, pancreatic cancer, brain tumors, dermatologic disease, macrocephaly, leukemia, lymphoma, or endometrial cancer. This is just one example. There are many different criteria and combinations of criteria that can lead to a genetic testing recommendation. Trying to remember all of these for just Hereditary Breast and Ovarian Cancer syndrome is challenging, but trying to remember all the possible patterns for all the different syndromes is near impossible unless it is your specialty, like medical genetics or genetic counseling.

In addition, there are common diseases for which there are hereditary variants that are not well appreciated in clinical practice. Examples of these are prostate cancer and diabetes mellitus. Prostate cancer is common and there is currently a great deal of controversy over screening for prostate cancer, yet it is still the second leading cause of death in men and not only does a family history indicate an increased risk at the familial level, there are several prostate cancer hereditary syndromes including Hereditary Breast and Ovarian Cancer, Lynch Syndrome, and Familial Prostate Cancer (which confusingly is actually a hereditary prostate cancer syndrome). In diabetes mellitus, there are clearly strong family histories for type II diabetes that are driven by familial level risk. This risk leaves the family susceptible to developing diabetes in the presence of cultural behaviors that promote its onset. However, there is also a hereditary syndrome called Maturity Onset Diabetes of the Young that is not well named as it can occur at any age. Maturity Onset Diabetes of the Young clinically presents as either type 1 or type 2 diabetes mellitus, but is not. Identifying those with this type of diabetes is critical as the treatment is often not the same and most individuals end up on the wrong treatment. In fact, some of them should not be on treatment at all. We don't yet know the prevalence of Maturity Onset Diabetes of the Young within the diabetic population, but it is higher than previously estimated. Risk calculators are being developed to help clinicians identify those whose diabetes may be a form of Maturity Onset Diabetes of the Young. The current version includes age of diagnosis, BMI at diagnosis, family history, and treatment regimens (https://www.diabetesgenes.org/mody-probability-calculator/).

To help, here are a few clinical pearls or "when to worry rules" you can keep in mind when seeing a patient[14,15]:

o Early age of onset. The earlier a person develops a health problem, the more likely it is to be hereditary

o Multiple involved relatives: Two or more close family members or two or more generations with the same health problem

o Disease occurring in the less often affected sex (e.g., males with breast cancer)

o Having related conditions in your family, such as having a family history of both diabetes and heart disease, or breast and ovarian cancer, or colon and uterine cancer

o Having unusually severe or recurrent disease in close family members

---

Defining risk levels for medical conditions

To exemplify the spectrum of conditions for which family health history is instrumental in determining risk and the level of risk conferred is provided in Table 1. The table is not a complete list; but rather meant to give a sense of the broad impact of family health history. Check marks indicate that family health history alone identifies that level of risk. Text is provided to also describe validated risk calculators where they exist. Remember all diseases have a population level risk so that is not included in the table.

As mentioned in the examples, there are guidelines for how to measure risk and how to manage patients at each risk level. In some cases, measurement is performed using validated risk calculators that incorporate family history with other risk factors and in others it is driven entirely by family history. These guidelines are risk-based, meaning that they tailor the management strategy to an individual's risk level. Highest risk? Get genetic testing. Lower risk? Maybe start screening earlier or with a different technology. Risk based guidelines were rare 20 years ago but have been on the rise. Some examples of current guidelines are presented in Table 2.

TABLE 1 Examples of familial and hereditary risk level and tools to measure each.

Disease	Familial risk level	Hereditary risk level
Breast cancer	BRCAPro, Gail, Tyrer-Cuzick	✓
Colon cancer	✓	✓
Ovarian cancer	✓	✓
Prostate cancer	Polygenic risk scores	✓
Pancreatic cancer		✓
Huntington's chorea		✓
Alzheimer's disease	✓	
Osteoporosis	✓	
Multiple sclerosis	✓	
Diabetes	Polygenic risk scores	✓
Sickle cell and thalessemia		✓
Cystic fibrosis		✓
Several types of cardiomyopathy		✓
Several types of arrhythimias		✓
Coronary artery disease	ASCVD and polygenic risk scores	
Familial hypercholesterolemia		✓
Poly cystic kidney disease		✓
Hypothyroidism	✓	
Pituitary cancer		✓
Glaucoma	✓	
Macular degeneration	✓	
Systemic lupus	✓	
Rheumatoid arthritis	✓	
End stage renal disease	✓	
Atrial fibrillation	Polygenic risk scores	
Early onset Parkinson's disease		✓
Migraines		✓
Epilepsy	✓	✓

TABLE 2 Examples of risk-based guidelines.

Organization	Guideline
National Comprehensive Cancer Network	Screening for Hereditary Breast and Ovarian Cancer based on family history only
National Comprehensive Cancer Network	Screening for Lynch syndrome based on family history only
American Cancer Society	Breast MRI for breast cancer surveillance in those with a lifetime risk $\geq 20\%$
American Cancer Society	Tamoxifen or raloxifene for breast cancer prevention in those with a 5-year risk $\geq 1.66\%$
American Diabetes Association	Screening for risk of Diabetes
American Heart Association	Initiating cholesterol treatment for those with a 10-year atherosclerotic heart disease risk $\geq 7.5\%$
Society of Vascular Surgery	Abdominal Aortic Aneurysm screening for those with a family history
American Cancer Society	Colonoscopies starting at 10 years younger than the age of colon cancer onset in a first degree relative
United States Preventive Services Task Force	Osteoporosis screening at earlier age (50–64) if a parent has had a hip fracture
National Institute for Health and Care Excellence	Screening for familial hypercholesterolemia
American Heart Association	Screening relatives of those with sudden cardiac death at age < 50

Clinical validity and utility of family health history

So how do we know if the family health history our patients provide is accurate? This question has existed since the emergence of our modern health care environment, oriented around the health system instead of a provider caring for whole families in their home. The fact is that medical conditions are complex, and getting more so, impeding the ability of the general public to accurately report the level of detail needed in a high-quality family health history. In addition, many individuals still prefer to keep their medical problems private, even from their relatives. Sharing of information is even more complicated in the setting of HIPAA privacy rules and the inability to share medical information between two related patients within a single health system. The gold standard for comparing family history data in the research setting is obtaining relative's medical record data, which is not feasible in routine clinical care at the moment (maybe one day in the not too distant future that will change!). While the clinical and research communities are working on these challenges (and some potential solutions will be suggested in later chapters), clinicians must go on providing clinical care as best we can, and right now the standard of care for family health history is self-report of relatives' medical histories during the office visit. We can certainly improve the quality of the data by: (1) educating patients about the value of family history; (2) giving them guidance on how and what to collect from their family members; and (3) allowing them time to gather the information. In fact, using online tools has been shown to significantly improve the amount and quality of family health history data reported, and in some cases has been comparable to data gathered by genetic counselors.[16]

Family health history's relationship to genetic testing

Throughout the first chapter and this one we have described how family history can be used to identify risk for disease. We've also touched on the levels of disease risk and that one of those levels is termed "hereditary" risk. In these individuals, the family history is indicative of a highly penetrant genetic variation that drives the disease incidence in the family. While this is the most common way for providers to use family health history there are other relationships between family health history and genetics/genomics. These will be described in more detail in later chapters, but I will quickly summarize here how family health history contributes to genetics:

1. identify the likelihood that a highly penetrant genetic variant is present
2. mediate the penetrance of a known pathogenic variant
3. provide evidence to clarify whether a variant of undetermined significance (VUS) is indeed significant or not
4. guide the search for new genetic drivers of hereditary disease

Identify the likelihood that a highly penetrant genetic variant is present

This is what this chapter has largely been devoted to describing. Who should undergo genetic testing and what genetic tests should be performed? We will go in to more detail about the process for testing in the clinical setting in later chapters. **The key takeaway for this chapter is that family health history is the most effective clinical tool available for identifying the hereditary risk group and allows for appropriate risk stratification of populations**. In fact, risk stratifying populations is one of the best examples of how "precision medicine" can not only improve an individual's health, but also improve the health of a population. With the advent of Accountable Care Organizations and the transition to value based payments, adopting a broad-based risk assessment strategy using family health history will become increasingly valuable.

Mediate the penetrance of a known pathogenic variant

This can take two forms. One is in hereditary syndromes, looking at highly penetrant variants, and the other is in the common chronic disease realm looking at polygenic risk scores.

For the highly penetrant variants. We already know that even with highly penetrant variants there is a wide range of disease penetrance. For example, with *BRCA1* mutations the lifetime risk of developing breast cancer is 50–85%. Why the range? Why isn't it just 85% or 100% for that matter? This variation in developing disease is called penetrance and we have reason to believe that the variation is due to both interactions between the identified gene and other genes, and between the identified gene and environmental/behavioral exposures (such as chest wall radiation or smoking). The problem with not being able to better refine the risk level is that I, as a patient, might view a preventative mastectomy more positively if my risk is 85%, but if it's 50% I might want to take a different course of action. Family health history can help refine likelihood of disease occurring. For example, with *PALB2*, the gene I mentioned earlier that contributes to hereditary breast cancer risk, the penetrance ranges from 30% to 60% for lifetime risk of breast cancer (Fig. 5). However, if you have no relatives with breast cancer your risk is 33% and if you have at least two with breast cancer by age 50 then your risk is 58%.[8] The difference between 33% and 58% is significant and may change the risk management options available to the individual.

For the polygenic risk scores. Polygenic risk scores will be discussed in more detail in later chapters but I introduced them briefly in the familial risk level discussion. These are those low risk variants called, single nucleotide polymorphisms (SNPs), that individually don't account for much but combined can confer a greater risk of disease, typically in the 2–3 times normal range. Polygenic risk scores only look at genomic results and not at environmental

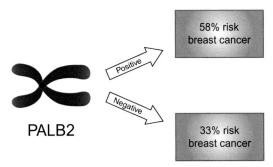

FIG. 5 How family health history can change the level of disease risk associated with a gene. The example of PALB2.

risk factors and do not incorporate family history. In comparing family health history risk level to polygenic risk score risk level, there is some overlap. Both methods of risk assessment find many of the same people, but each also identifies individuals not identified by the other. This suggests that the two methods are complementary and should be used together. Much more research in this area is warranted, but as we learn more about the genome, the polygenic risk scores may start to perform better than they do now. Even so, they cannot account for the interactions between genes and environment and there will likely always be role for family health history in interpreting polygenic risk score results.

Provide evidence to clarify whether a variant of undetermined significance (VUS) is indeed significant or not

Genetic testing is unfortunately not as clear cut as we would like it to be. Currently, there are about 6000 known single-gene genetic disorders and more than 250,000 known pathogenic variants; but this list changes every day. One report suggests that we identify 17,000 new pathogenic variants every year.[17] Managing this ever-growing list is in and of itself an enormous undertaking, but when you throw in the variants of undetermined significance (VUS) it becomes and almost insurmountable task. Variants of undetermined significance are common. They arise when we compare an individual's gene sequence to what we call the reference genome. The reference genome is a combination of many different healthy individuals' DNA, because there is so much variation in the human population. This variation arises from an ongoing mutation rate in the population. Every time a baby is born there are 44–82 new variants in the baby's genome compared to their parents, due to copying errors during development.[18] Most of the time these are not in coding areas of the genome and even the ones that are typically do not impeded the function of the gene. However, this means there is no single "normal" genome to compare our patient's results to. Instead, the reference genome is a database containing many healthy individuals' genomes representing the spectrum of normal at each nucleotide position. But there are two problems with the reference genome: (1) we don't yet know the spectrum of normal at each position (particularly for those who are not Caucasian as they are underrepresented in research and genomic studies); and (2) there are still gaps in the reference genome—places where we don't know anything about the DNA sequence. We are closing these gaps, but until we know all the possible normal options at every position, we will continue to have to deal with VUSs. A variant is called a VUS when we are unsure if it could be disease causing or normal (Fig. 6). Often, they are in areas where it could potentially affect the function of a gene, but we don't have data to clearly say one way or another. Sometimes animal models can help us figure this out. By inserting a VUS variant into an animal genome we can look for evidence of nonfunctional proteins or development of disease in the animal. However, that

VARIANT CALLS

FIG. 6 Variants of undetermined significance (VUS) means exactly that—unsure of the significance.

takes time and a source of funding to investigate. In the meantime, there is one source of information that is readily available that can provide insight into the pathogenicity of the variant—family health history. If there is a strong family history of a disease that correlates with the type of dysfunction that should arise given the function of the affected gene, then we have evidence to suggest the VUS is in fact meaningful. Further evidence to support the link between the VUS and disease could be obtained by testing relatives with and without the disease to see if those with disease have the variant and those without do not. If this pattern holds, it is strong evidence that the VUS is indeed pathogenic (see Chapter 4 for more on VUS).

Guide the search for new genetic drivers of hereditary disease

Families with a large number of relatives with disease or who have many individuals develop disease at a young age are increasingly used to help identify new single gene variants. As alluded to with VUSs, a strong family health history can help solidify the link between a variant and disease. They can also be used to explore the underlying genetic drivers of disease. A dramatic example of this is Mary Claire King's discovery of the *BRCA1* gene in 1990,[19] when I was just graduating from college. She identified the gene by doing linkage analysis from families with a family history that suggested an autosomal dominant breast cancer pattern. This method of discovery continues to be successful today for a wide-ranging number of conditions. While what I am describing is a scientific research method and is not immediately useful in clinical decision making, the results have longstanding implications for clinical care. In addition, the idea of establishing registries or large databases of family histories that could be used to identify potential cohorts for research studies could begin with collecting those family histories in the medical setting.

Summary and conclusions

In summary family health history is one of the most critical tools a clinician can have to assess disease risk in their patients. It is unique in that it provides information on both genetic and environmental contributors to disease and is currently a more complete resource for understanding risk than genetics is. For diseases that we know have strong genetic drivers, such as hereditary syndromes, it is an effective method for identifying those who need genetic testing. In addition, these syndromes are no longer considered to be truly rare and together they comprise a significant proportion of the general population. Beyond identifying hereditary risk, family history can also help identify those at familial level risk who would warrant more intensive risk management strategies. Lastly, it can also help to guide the interpretation and discovery of genetic variants.

Lori A. Orlando is the lead author for this chapter. First person statements are from her point of view.

References

1. Orlando LA, Henrich V, Hauser ER, Wilson C, Ginsburg GS. The genomic medicine model: an integrated approach to implementation of family health history in primary care. *Pers Med.* 2013;10(3):295–306.

2. Orlando LA, Wu RR, Beadles C, et al. Implementing family health history risk stratification in primary care: Impact of guideline criteria on populations and resource demand. *Am J Med Genet C Semin Med Genet.* 2014;166(1):24–33. https://doi.org/10.1002/ajmg.c.31388.

3. Bennett RL. The family medical history. *Prim Care.* 2004;31(3):479–495. vii-viii, https://doi.org/10.1016/j.pop.2004.05.004.

4. Ziogas A, Horick NK, Kinney AY, et al. Clinically relevant changes in family history of cancer over time. *JAMA.* 2011;306(2):172–178.

5. Beadles CA, Ryanne Wu R, Himmel T, et al. Providing patient education: impact on quantity and quality of family health history collection. *Fam Cancer.* 2014;13(2):325–332. https://doi.org/10.1007/s10689-014-9701-z.

6. Bao W, Hu FB, Rong S, et al. Predicting risk of type 2 diabetes mellitus with genetic risk models on the basis of established genome-wide association markers: a systematic review. *Am J Epidemiol.* 2013;178(8):1197–1207. https://doi.org/10.1093/aje/kwt123.

7. Mehrgou A, Akouchekian M. The importance of BRCA1 and BRCA2 genes mutations in breast cancer development. *Med J Islam Repub Iran.* 2016;30:369.

8. Antoniou AC, Casadei S, Heikkinen T, et al. Breast-cancer risk in families with mutations in PALB2. *N Engl J Med.* 2014;371(6):497–506. https://doi.org/10.1056/NEJMoa1400382.

9. O'Neill SM, Rubinstein WS, Wang C, et al. Familial risk for common diseases in primary care: the Family Healthware Impact Trial. *Am J Prev Med.* 2009;36(6):506–514. https://doi.org/10.1016/j.amepre.2009.03.002.

10. Qureshi N, Armstrong S, Dhiman P, et al. Effect of adding systematic family history enquiry to cardiovascular disease risk assessment in primary care: a matched-pair, cluster randomized trial. *Ann Intern Med.* 2012;156(4):253–262. https://doi.org/10.7326/0003-4819-156-4-201202210-00002.

11. Hampel H, Bennett RL, Buchanan A, Pearlman R, Wiesner GL. A practice guideline from the American College of Medical Genetics and Genomics and the National Society of Genetic Counselors: referral indications for cancer predisposition assessment. *Genet Med.* 2014. https://doi.org/10.1038/gim.2014.147.

12. Vogel VG, Costantino JP, Wickerham DL, et al. Effects of tamoxifen vs raloxifene on the risk of developing invasive breast cancer and other disease outcomes: the NSABP study of Tamoxifen and Raloxifene (STAR) P-2 trial. *JAMA.* 2006;295(23):2727–2741. https://doi.org/10.1001/jama.295.23.joc60074.

13. Gail MH, Brinton LA, Byar DP, et al. Projecting individualized probabilities of developing breast cancer for white females who are being examined annually. *J Natl Cancer Inst.* 1989;81(24):1879–1886. Online risk calculator available at: http://www.cancer.gov/bcrisktool/.

14. Reid G, Emery J. Chronic disease prevention in general practice – applying the family history. *Aust Fam Physician*. 2006;35(11):879–882. [884-875].

15. Yoon PW, Scheuner MT, Peterson-Oehlke KL, Gwinn M, Faucett A, Khoury MJ. Can family history be used as a tool for public health and preventive medicine? *Genet Med*. 2002;4(4):304–310. https://doi.org/10.1097/00125817-200207000-00009.

16. Murray MF, Giovanni MA, Klinger E, et al. Comparing electronic health record portals to obtain patient-entered family health history in primary care. *J Gen Intern Med*. 2013;28(12):1558–1564. https://doi.org/10.1007/s11606-013-2442-0.

17. Stenson PD, Mort M, Ball EV, et al. The human gene mutation database: Towards a comprehensive repository of inherited mutation data for medical research, genetic diagnosis and next-generation sequencing studies. *Hum Genet*. 2017;136(6):665–677. https://doi.org/10.1007/s00439-017-1779-6.

18. Acuna-Hidalgo R, Veltman JA, Hoischen A. New insights into the generation and role of de novo mutations in health and disease. *Genome Biol*. 2016;17(1):241. https://doi.org/10.1186/s13059-016-1110-1.

19. Hall JM, Lee MK, Newman B, et al. Linkage of early-onset familial breast cancer to chromosome 17q21. *Science*. 1990;250(4988):1684–1689.

Chapter 3

The connection between genetic variation, family health history, and disease risk

Lori A. Orlando, Vincent C. Henrich, and Brian H. Shirts

- A genomics primer: What are DNA variants.
- Why do we have single nucleotide polymorphisms (SNPs).
- Most SNP variants have no known medical relevance.
- Some SNPs are associated with disease.
- Family health history's relationship to genetic variants.
- Types of genomic tests currently available.
- What to expect when you get your genetic test result back.
- What to make of direct to consumer (DTC) genetic tests.

As noted in the previous chapter, genetics and family health history are connected through the genotype and phenotype relationship, which as described in Chapter 1, is complex and nuanced. For most of us, our only exposure to genetics was high school science class, where we learned about Mendelian inheritance. While Mendel's laws of genetic inheritance still apply today, the pathway between genotype and phenotype is mediated by a broad range of cellular events that are in turn mediated by other genes and/or environmental exposures. Many of these pathways remain unknown and will take decades to fully understand. This chapter will describe why we have genetic variants, why they are transmitted through populations, and how they can impact health and disease. In addition, we will delve into what genetic testing options are out there, the types of results you might receive, how to interpret the results, and what their limitations are.

A genomics primer: What are DNA variants

The Human Genome Project, an effort to systematically catalogue the sequence, location, and structure of human genes, completed sequencing the entire human genome in 2003. As a result, we now have a template for understanding the relationship between genomics, and health and disease. This effort also spawned related work in areas such as RNA (transcriptomics), proteins (proteomics),

Managing Health in the Genomic Era. https://doi.org/10.1016/B978-0-12-816015-2.00003-1

molecules produced by cells (metabolomics), the community of microbes that populate our bodies (microbiome), and the microbiome's DNA (metagenomics). While these are promising fields of research, they are rapidly evolving and too early to further describe in this book.

In the follow up to the initial assembly of the human genome, attention focused on the ~88 million variants scattered throughout the genomes of 2500 individuals who were selected from populations dispersed around the world. Some were within the protein coding regions of genes, but many were, surprisingly, in intergenic (between gene) regions. Most involved the substitution of a single nucleotide (A, C, G, or T) for another nucleotide, though occasionally there were small deletions or insertions (indels) that changed the number of nucleotides present in the gene (Fig. 1). Larger structural variants occurred less frequently. Could these seemingly minor variants explain why healthy people possessed a variety of traits, tendencies, disease susceptibilities, and lifespans? Based on the compiled sequence information from the 2500, a consensus human genome arose, called the reference genome (Fig. 2). With the reference genome in place, it became clear that every genome contains some (but not all) of these 88 million single nucleotide variants (SNVs) found in the analysis. About 64 million are "rare," that is, they are found in less than 1% of the population, 12 million are less rare (0.5–5% of the population), and 8 million are common (>5% of the population) (Fig. 3). Additionally, every human carries a number of variants that do not exist in the 2500 individuals whose genomes were analyzed. *Single nucleotide polymorphism (SNPs) are single nucleotide variants (SNVs) that occur in at least 1% of the population.*

An average genome includes 4–5 million SNVs from the reference genome. Where did these variants come from? With every birth there are random errors

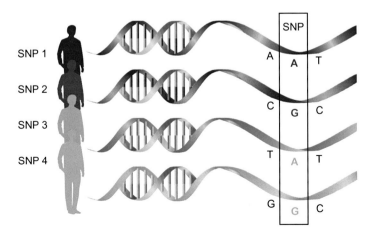

POLYGENIC RISK SCORE = SNP 1 (T) + SNP 2 (G) + SNP 3 (A) + SNP 4 (C) = 2.1

FIG. 1 Single nucleotide polymorphisms.

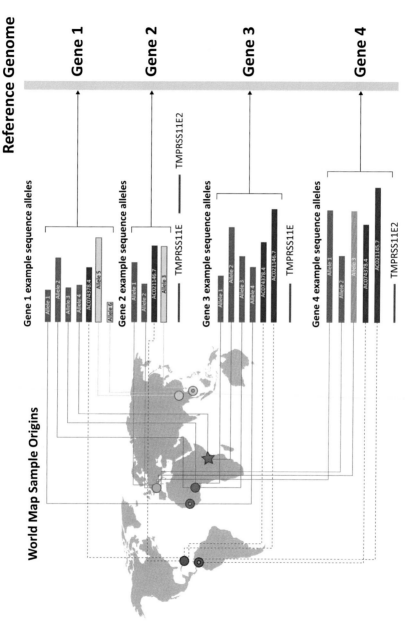

FIG. 2 The consensus human reference genome.

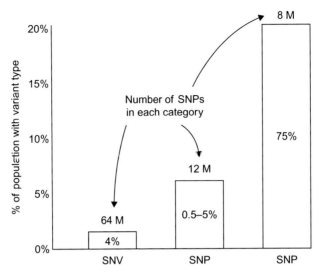

FIG. 3 Single nucleotide variants (SNVs) vs. single nucleotide polymorphisms (SNPs) and their prevalence in the population.

in copying DNA that results in an average of 60 new variants. As long as these are not deleterious, they are incorporated into the gene pool along with all the prior generations' variants. So, going back to the 4–5 million SNVs, each arose randomly during a birth at some point in our ancestral past. The vast majority of these are common enough to be classified as SNPs, though we each also harbor between 40,000 and 200,000 rare SNVs, whose distribution in the human population is much more restricted. Most of these "rare" variants arose in a recent ancestor, in the last few centuries, and have not had time to diffuse beyond our extended families. *These variants are considered "family-specific" and will be discussed in more detail in Chapter 4.* The impact of a SNP is determined by its location (in a gene or gene regulatory area), and whether the nucleotide change results in a change to the gene's protein's function or not. The question is: do variants like these raise the risk for disease?

Why do we have SNPs?

SNPs are variants that arose in an ancient human ancestor. As such, they have had plenty of time to spread throughout the human population, significantly more time than the family specific SNVs described earlier. However, their frequency is not uniform. Instead, the frequency varies depending upon how long ago the SNP arose. The more recent, the less time it has had to spread. The migration out of Africa can be mapped using SNPs as both an ancestral geographic and time marker (Fig. 4). For example, SNPs seen in Asian populations; but not European ones, are more recent variants, and their presence can

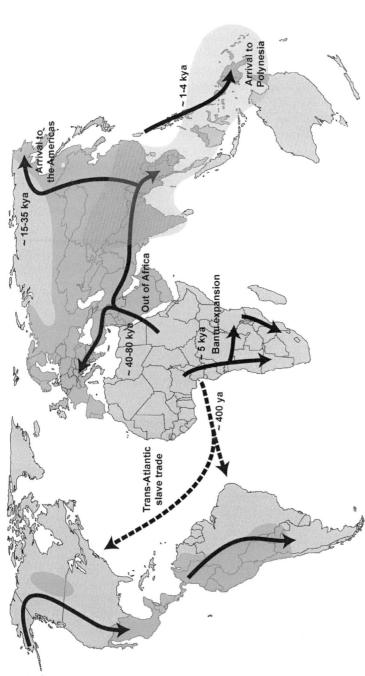

FIG. 4 The migration out of Africa as measured using SNPs. Major human migrations across the globe and presence of archaic ancestry in the genomes of modern human populations. *Red arrows* indicate the major migrations of *Homo sapiens* to colonize the world after the out of Africa exodus. *Blue arrows* indicate some more recent migratory events (<5000 years ago), and *dashed arrows* represent the historical migrations related to the Trans-Atlantic slave trade. Approximate geographic areas of modern human populations presenting Neanderthal or Denisovan ancestry are shaded in *blue* and *orange*, respectively, according to the Simons Genome Diversity Project. This map does not present a comprehensive evaluation of the populations presenting archaic ancestry, as only populations for which genetic profiles are available are plotted. The Neanderthal ancestry observed in American populations does not reflect in situ admixture with Neanderthals but, instead, their varying levels of European ancestry.

be traced down to their native American descendants (from the Bering strait migration). In addition, if a population becomes isolated for long enough, they can acquire their own unique SNPs that are not present in the general population. These polymorphisms can be quite old, but are still "restricted" to a specific population. Even if the population did not remain completely restricted over time, if they remained isolated long enough to develop their own unique SNPs, and then began mingling with other populations again, the frequency of the variant will form a clear pattern pointing to the "founder" population (the circle with the highest frequency at the center, Fig. 5). Factor V Leyden is a perfect example of a polymorphism with a founder population. It is present in ~5% of northern Europeans, but has a steadily declining prevalence in populations that live further away from this epicenter, until finally becoming nonexistent in east Asia.

Other factors that influenced the proliferation and frequency of SNPs include the founder population's rate of expansion. Rapid population expansion,

FIG. 5 Example of how a founder variant can spread within a geographic area.

migration, and intermingling between population groups (i.e., admixture) increases the frequency of a variant and broadens its distribution. Selection also plays a role. When a SNP confers a health advantage, it is more likely to propagate, since individuals with the SNP preferentially survive. An example is sickle cell anemia. Sickle cell anemia is caused by a SNP that has been traced to three different founders in tropical Africa. Why, many have asked, would such a terrible disease continue to be transmitted through the population, when sickle cell anemia is highly lethal among adolescents, who are too young to reproduce and pass on the polymorphism? The answer is that it confers resistance to malaria, also a highly lethal disease affecting those living in tropical Africa, and sickle cell anemia is limited to those who inherit the SNP on both strands of their DNA (i.e., they inherited it from both their mother and their father). Further support for its selective survival advantage includes, the SNP spontaneous arose in three different founders, and its presence in southern Europe where malaria was also, historically, a significant cause of death, but not in northern Europe where it was not.

So what happens, genetically, if a population remains relatively isolated over many generations? Iceland offers some useful insights. About 90% of Iceland's current population is a descendant of a small group that settled on the island beginning in 874 CE. Despite the relative isolation of the population for thousands of years, Icelanders are not significantly different from other populations, except that the population carries a number of SNPs that are not found in other populations and a few of them are medically relevant.

The variation of SNPs across populations (and sub-population) creates significant challenges in interpreting the human genome. Challenges that are exacerbated by the strong Eurocentric bias present in most genomic studies. How do we know if a SNP is common or not, disease causing or not, if the reference population is different from the population being tested? An example of how failure to incorporate diverse populations in genomic studies can lead to misclassification and potential harm was published in the New England Journal of Medicine in 2016. Early genomic association studies of Caucasian populations linked a SNP with hypertrophic cardiomyopathy, ultimately classifying it as pathogenic. Reanalysis in African American populations found the SNP to be a common and benign polymorphism, but not until after many had been told they carried a pathogenic variant for hypertrophic cardiomyopathy.[1] When analyzing diverse genomes, from people across the planet, we find tens of millions of points in the DNA that were altered early enough in human history to be detected repeatedly in present day human descendants and are normal variants—hence the reference genome contains many options for normal variation at many sites. The rarity of a founder mutation inherited from an ancestor who lived a few thousand years ago in a small human population explains why most SNPs and SNVs have only a single alternative base variant. In fact, the rare instances where two SNPs occur at the same base point, are most likely due to two founder events occurring in the same position.

Most SNP variants have no known medical relevance

To assess the relationship between SNPs and disease, researchers have employed an analytic method called genome-wide association studies (GWAS), which compare the frequency of diseases in those with one SNP variant to those with the alternative variant (Fig. 6). Over the years, GWAS has implicated numerous SNPs in disease processes (Table 1), though the relationships are not strong, and many were inconclusive. In fact, most SNPs have no known association with disease. The problems with GWAS are largely due to poor study design. For example, keeping in mind Dr. Henrich's family case of stroke from Chapter 1, a GWAS to find a SNP associated with stroke that did not take into consideration the type of stroke, triggering conditions, brain localization, age of onset, and incidence in the population would fail to find a meaningful connection. Another problem is that SNPs have different frequencies in different populations. Controlling for ancestral background is imperative to avoid misinterpretation of a GWAS association, as was described in the hypertrophic cardiomyopathy example.

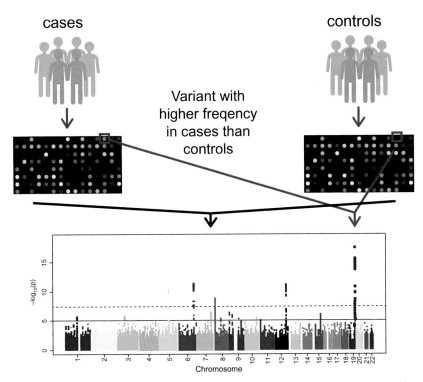

FIG. 6 Genome wide association studies (GWAS) to identify the relationship between a variant and disease.

TABLE 1 Examples of common single nucleotide polymorphisms (SNPs).

Gene	Normal role	Effect of SNP(s)	Cellular effect	Medical impact	Frequency	SNP identity and comments
APOE	Blood lipoprotein carrier	Alters an amino acid in ancestral (APOE*e4) protein	Reduces affinity for low density lipoprotein receptor (LDL-R)	Reduces e4 risk of coronary heart disease (CHD)	≥50% in developed countries	rs387906567
		Alters an amino acid in APOE*e3 variant	Reduces affinity for LDL-cholesterol	Reduces e3 risk of coronary heart disease (CHD)	≤10%	rs429358
Factor V	Regulates blood clotting process	Changes amino acid	Increases blood clotting rate	Increase risk of thrombophilia	1–5% in European populations	rs4524. Leiden polymorphism
HOXB13	Regulates cell cycle by affecting activity of specific target genes	Changes single amino acid in HOXB13 protein	Increases rate of proliferation in prostate cells	Increases risk of aggressive prostate cancer	0.1–0.3% in several populations	rs138213197
BRCA1	Repairs damaged DNA in cell	Alters sequence of noncoding mRNA region and disrupts mRNA stability	Increases susceptibility to loss of DNA repair function in cells	Increases risk of breast/ovarian and other specific cancers	≥1% in Askenazi Jewish descendants	Am J Hum Genet. 1997; 60: 505–514, 23andMe tests, rs80357720
BRCA2	Repairs damaged DNA in cell	Alters sequence of noncoding mRNA region and disrupts mRNA stability	Increases susceptibility to loss of DNA repair function in cells	Increases risk of breast/ovarian and other specific cancers	≥1% in Askenazi Jewish descendants	Am J Hum Genet. 1997; 60: 505–514, 23andMe test, rs80359550

Continued

TABLE 1 Examples of common single nucleotide polymorphisms (SNPs)—cont'd

Gene	Normal role	Effect of SNP(s)	Cellular effect	Medical impact	Frequency	SNP identity and comments
MSH2	Repairs sequence errors occurring during DNA replication	Alters a single amino acid in MSH2 protein sequence	Increase susceptibility to loss of DNA mismatch repair	Increases risk of Lynch syndrome cancers	No data. Found independently in unrelated patients	rs63750828
MSH2	Repairs sequence errors occurring during DNA replication	Truncates MSH2 protein sequence	Increase susceptibility to loss of DNA mismatch repair	Increases risk of Lynch syndrome cancers	No data. Found independently in unrelated patients	rs587779063
PPARG	Regulates cellular regulation of fatty acid metabolism	Alters a single amino acid in PPARG protein sequence	Complex effect on lipid metabolism in cells	Reduces risk of type 2 diabetes, Elevates diet-dependent BMI	No data	rs1805192

Some SNPs are associated with disease

Early genetic studies were successful in identifying SNPs associated with disease by focusing on families with diseases that had clear Mendelian inheritance patterns. For example, the variant responsible for Huntington's disease is an inherited autosomal dominant mutation with complete penetrance (one of the very few pathogenic mutations which always *cause* disease). Thus, the pattern was easy to identify and the disease easy to follow in affected families. The causative variant alters the structure of a protein that is able to function normally during neural development; but becomes entangled with adult neural cells promoting neural degeneration in adulthood. Another example is *BRCA1*, the poster child for adult genetic testing. Mary Claire King discovered the gene when she was given permission to add questions about family members with breast cancer to a nationwide survey. Through the survey she was able to identify a large number of families with an exceptionally high incidence of breast cancer. Genetic analyses of the families lead her to the gene. Interestingly, BRCA1 and BRCA2 are responsible for repairing damaged DNA. SNPs that alter the function of the BRCA1 and BRCA2 proteins, prevent DNA from being repaired, leading to an accumulation of deleterious mutations, and ultimately cancer. Nevertheless, some carriers of pathogenic *BRCA1* or *BRCA2* variants manage to live a cancer-free life suggesting that there's more to the story than solely genetics.

Summary of SNPs

- SNPs, like all genetic variation, are the result of random mutations.
- SNPs cannot be associated with diseases that decrease child bearing (e.g., they die prior to reaching their child bearing years, or they are unable to bear children).
- SNPs can confer a selective health benefit, particularly if it enhances the likelihood of having offspring.
- In order for a SNP to become prevalent enough to be detected in present day populations, it had to arise in an ancestor (the founder) several thousand years ago.
- The frequency and distribution of a SNP depends in part on the founder population. A heterogeneous distribution can occur when a SNP arises in a relatively isolated population and slowly spreads to nearby subpopulations.
- Most SNPs fulfill no discernible function, but can serve as sign posts for other medically relevant variants, if they are located in physical proximity to each other.
- The NIH maintains a SNP database that contains location, function, frequency, and research publications (www.ncbi.nlm.nih.gov/snps).

Family health history's relationship to genetic variants

The four relationships between family health history and genetic testing were described in Chapter 2. Understanding these relationships is crucial for effectively using and interpreting genetic test results. As a quick reminder there are 3 levels of disease risk that are closely correlated with genetic variants.

The first is population level risk where the disease risk is no greater than the general population's, the second is familial risk that increases the risk of disease above the level of the general population, but not as high as the third, hereditary risk (Fig. 7). In both the familial and hereditary levels of risk, the influence of genes on risk is mediated by environmental and (other) genetic factors. Unfortunately, we know very little about how gene-gene interactions affect disease risk. For this reason, we will focus the discussion on gene-environment interactions though we still have a lot to learn about these as well. Variants associated with hereditary conditions are monogenic (i.e., single gene mutations) that alone confer a very high level of risk. In these cases, environmental interactions can still adjust the risk up or down, but the impact is relatively small compared to that conferred by the gene. Conversely, variants associated with familial risk have lower impact on disease risk and environmental interactions can adjust it to a greater degree. With that in mind we will return to our discussion of family health history-gene relationships: (1) family health history determines who to perform genetic testing on, (2) family health history mediates the risk of developing disease, (3) family health history informs whether a variant of undetermined significance might be pathogenic or benign, and (4) family health history helps identify new pathogenic variants.

All of these relationships hinge on two important factors- the first is that, as previously described, family health history represents the impact of both shared genes and shared environment on disease risk in the family, and second that genetically driven diseases frequently have incomplete penetrance. There are only a few adult onset conditions, such as Huntington's, that are completely penetrant, causing all individuals with the variant to acquire the disease. In most cases there is variability in the likelihood of developing disease and thus penetrance is incomplete. *BRCA1* and *BRCA2* variants are an example of this

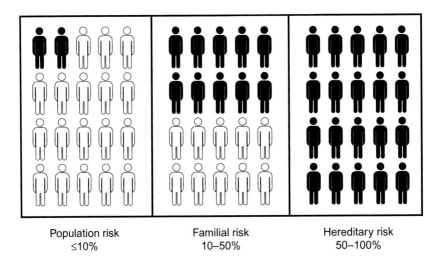

| Population risk | Familial risk | Hereditary risk |
| ≤10% | 10–50% | 50–100% |

FIG. 7 The three levels of disease risk: population, familial, and hereditary.

concept. Adults with a pathogenic variant in *BRCA1* or *BRCA2* have a life-time risk of breast cancer as high as 80%. Thus more than 20% of *BRCA1* or *BRCA2* carriers do not develop breast cancer. As mentioned above, much of the variability in developing disease is presumed to be due to gene-gene and gene-environment interaction; but the mechanisms vary for different diseases and genes. Chapters 5 and 6 will explore these concepts in more detail with supporting cases. Since family health history also reflects gene-environment interactions, family health history collection is an excellent method for ascertaining risk reflected by incomplete penetrance. In these scenarios, a strong family health history can both inform who to perform genetic testing on and how high the penetrance may be (i.e., the level of risk conferred by the variant).

But what about the scenario when a family has a preponderance of family members affected by a disease, but genetic testing does not identify a pathogenic variant? In this case, there are two options. If the non-pathogenic variant is a variant of undetermined significance AND the gene has a function that is plausibly related to the disease, then the family information may change the variant's status from undetermined to pathogenic (see Chapter 4). If there are no pathogenic or undetermined variants identified then there are two possibilities. The first is that a pathogenic variant exists that has yet to be discovered, and the second is that the disease is primarily driven by the shared environment. This scenario is not surprising when one considers that nothing is known about 90% of the 22,000 genes catalogued by the Human Genome Project. Either way, **the lack of an identified genetic driver, does not change the reality for the individuals in the family—they have an increased risk for the disease**. As such they should continue to be offered risk reducing options based upon their familial level of risk. For example, if a women's lifetime risk for breast cancer is > 20%, even if her genetic testing is unrevealing, she should be offered breast MRI as an adjunct to mammography for breast cancer screening starting at the age of 35.

Types of genomic tests currently available

In today's medical practice we are inundated with genetic test options, which seem to be expanding by the day. I routinely get emails from new companies offering genetic tests. How do I know if I can rely on these tests? How about the company? What about insurance coverage? There are so many questions in this underdeveloped field that it is hard for the non-expert to navigate. Our purpose in this discussion is not to turn the reader into an expert, but to give you a framework with which to understand the different offerings.

Broadly there are five different types of genomic tests: single gene variant testing, genotyping, genotype panels, single gene sequencing, and whole genome or exome sequencing.

- Single gene variant testing is the oldest and simplest type of test available. In this method a single gene is analyzed for known pathogenic variants and their presence or absence is recorded. It does not analyze the entire gene, only those variants that have been validated as disease causing. This test is only useful when a single gene is known to be the sole cause for the disease. One benefit is that the test is quick and relatively inexpensive. In addition, because only known variants are analyzed, there is no extraneous information to potentially confuse the provider or patient. However, because only variants known at the time of the testing are analyzed, when new variants are discovered, a whole new test (at an additional cost) must be run. In the current era, with our rapidly expanding knowledgebase, these tests can quickly become outdated.

- Genotyping is the same as single gene variant testing but analyzes the genome rather than a single gene. It has similar pros and cons to single gene testing; but does have broader coverage over multiple genes associated with a disease and is more commonly used today than single gene single variant testing.
- Genotyping panels are the same as genotyping, but instead of genotyping genes associated with a single disease, they include large numbers of unrelated genes that cover a broad array of diseases. Some panels can include hundreds of genes covering 20 or more diseases. Panels can be an efficient and cost-effective method for answering a specific clinical question, while simultaneously screening for a number of other conditions. The downside of panels is that many insurance companies will not reimburse for panels, and it is not always clear what to do with the results from the "other" genes.
- Single gene sequencing sequences every base pair in the gene. In situations where a gene has an established causal relationship with a disease, it may be more effective to sequence the whole gene rather than perform a genotype. This is particularly, helpful when a gene may contain many pathogenic SNPs, or could have a novel variant that renders the protein non-functional. When sequencing a gene, every nucleotide in the gene is analyzed and recorded, thus, as new pathogenic variants in the gene are identified, the sequence can be re-analyzed and updated without having to re-run the test.
- Whole exome and whole genome sequencing are the last type of genomic test. In whole genome, the entire genome is sequenced, while in whole exome, only the protein coding regions of the genome (the exome) are sequenced. These tests give almost complete information about an individual's DNA and are thus relatively expensive and time consuming to perform. In addition, since we don't know anything about 99% of the genome, they yield large amounts of data we don't currently have a medical use for. The benefit of whole genomes or exomes, is similar to single gene sequencing, you have all the information available to you for future reanalysis as new information becomes available. A potential downside, is what to do with pathogenic findings that are unrelated to the reason the test was performed. Both whole genome and whole exome sequencing have been very helpful in sick patients, often children, who clearly have an underlying medical disease but the etiology remains an enigma even after extensive evaluation.

What to expect when you get your genetic test result back

When you receive your genetic test results you will be given information on the type of test, what was analyzed, and what was found. Each company and lab report this information in different formats, with different content. Thus, results (and often interpretation) are not standardized across labs. There is, however, a standard framework for classifying variants that has been widely adopted. Variant classification guidelines, proposed by the International Association for Cancer Research (IARC) in 2008,[2] were designed to classify variants associated with monogenic diseases in genes known to be associated with a disease (phenotype), not for those identified via population screening or in genes without a clear connection to a disease phenotype. This distinction is important, because in genes known to cause a disease, variants can either cause the disease or not, and thus are simply designated as pathogenic (disease causing) or benign (not disease causing). However, when a variant is seen for the first time or has not been well studied, it is not always clear if the variant causes disease. Hence, variants in these situations are classified as pathogenic, likely-pathogenic, variant of undetermined significance, likely-benign, or benign (Fig. 8). The middle three classifications do not denote separate biological phenomena, but rather levels of evidence, which is why a strong family health history can change a variant's classification from undetermined significance to likely pathogenic if the gene can be plausibly associated with the disease (phenotype). The IARC framework quantifies how much evidence is needed for each classification level: if combined evidence indicates a 99% chance that a variant causes the disease, this is sufficient to classify it as pathogenic, if it is 99–95% chance it is classified as likely pathogenic, with a 5–95% chance it is classified as a variant of undetermined significance, a 1–5% chance of causality classifies it as likely benign, and <0.1% chance of causality classifies it as benign. Bayesian methods are employed to assess how much new evidence changes the likelihood that the variant is disease causing. Once there is enough evidence to cross one of the thresholds, the variant's classification is changed. Note that a report with a variant of undetermined significance (VUS) does _not_ suggest intermediate risk, but rather the lack of evidence (or conflicting evidence) about the variant's likelihood to cause disease.

The more recent ACMG-AMP guidelines for variant classification are the ones most commonly cited at the time of this book.[3] Their underlying 5 tier framework for variant classification is the same as that in the IARC framework; however, Bayesian analysis is not used for evidence assimilation. Instead the

VARIANT CALLS

FIG. 8 Variants of undetermined significance (VUS) means exactly that—unsure of the significance.

guidelines list several categories of data that can contribute varying levels of evidence to variant classification, and describes how these categories should be combined. Data categories include: population data, computational and predictive data, functional data, family segregation data, de novo data, allelic data, database information, and other (see Chapter 4 for a more complete discussion of these types of information). Examples of how these are used are presented in Chapter 4. Since there remains flexibility in how these rules are applied, differences in classification between clinical laboratories can occur—explaining some of the discrepancies in reports between labs. Because clinical judgment and expertise are necessary to apply these rules, it is recommended that classification and interpretation be limited to molecular pathologists and medical geneticists experienced in variant assessment.

These 5-tier categories were developed with the most common dominant and recessive diseases in mind. There are some variant classes that are important for medical management that do not fit into the 5-tier category. These include pharmacogenetic variants, which usually only have an effect on health in the context of drug treatment, and risk-modifier variants, whose effect is too small to justify changing medical management when considered independently.

In summary, when receiving the results of a genetic test, the report will contain information about the variant and its classification based on its likelihood to cause disease. Given the nuances of classifying variants, in some cases different labs may classify a variant differently. Most importantly, remember a *variant of undetermined significance means only that there is not enough evidence to determine one way or another whether it may be disease causing or not.*

What to make of direct to consumer (DTC) genetic tests

With increasing coverage of genomics in the press, the public's awareness about disease risk is also increasing. A turning point was Angeline Jolie's public statement about her family's history of breast and ovarian cancer, and her pathogenic *BRCA1* variant. The subsequent furor that followed, led to increased genomic testing and greater interest in assessing a family's medical conditions. Unfortunately, medical science trailed the demand, in part because the concepts underlying Ms. Jolie's revelations were complex and not as simple as the lay press represented. This disconnect created an opportunity for direct to consumer (DTC) genomic testing companies to market to those who wanted to have a genetic test, but whose medical provider was not involved in the decision making.

Initially DTC companies offered a genotyping (thus SNP-based) panel that represented a wide variety of both medically relevant and "interesting" but not medical relevant phenotypes (e.g., ear wax type, ancestry, athleticism, etc.). The quagmire that ensued regarding the medically relevant SNPs, prompted the FDA to regulate DTC testing. Regulations initially prevented testing for any disease-causing SNP without the direct involvement of a medical provider. Over time the FDA has loosened their regulations, allowing testing of selected disease-causing SNPs, such as those in *BRCA1* and *BRCA2*, without a provider's order.

The case that follows highlights some of the concerns that arise with DTC testing for disease risk. A friend's 66-year-old husband was diagnosed with an aggressive form of prostate cancer. She was seeking insight about the "cause" of his cancer, and suspected her husband's exposure to hazardous chemicals at work might have been the culprit. To better understand his prostate cancer, her husband underwent genomic testing at a DTC company. It is important to note that at this point he had not gathered any family health history information—he had jumped straight to the DTC test. The report left the couple believing that he carried no cancer variants and thus that the environmental hazards encountered in his work were the reason for his aggressive cancer. Afterwards, an assessment of his family health history revealed a mother and several maternal relatives with colorectal cancer. There is at least one syndrome that includes both colorectal cancer and prostate cancer, and thus the possibility of a genetic connection is plausible and should be explored. This highlights a few important concepts: (1) awareness of his family health history information would have been useful before his cancer onset—if there had been greater awareness of its potential clinical value, (2) even a thorough genomic analysis probably would not by itself have led to a pre-cancerous risk assessment that was not already apparent from his family's health history, and (3) current DTC testing cannot, in most cases, produce informative insights about a genetic connection to cancer, since they are unable to test for these. The restricted number of SNPs that could be evaluated on the DTC test was not clearly understood by the couple and confused them into believing his cancer was unquestioningly "environmental." Unfortunately, most consumers believe that all genomic tests look at everything in their entire genome and given them "THE" answer. Few are sophisticated enough to thoughtfully evaluate the different testing options and compare their pros and cons.

Another complication of DTC services is that while they were designed to fill the gap created by medical providers' unwillingness to journey into wholesale genetic testing for disease risk, in the absence of clear guidelines and evidence, the result has been that patients obtain DTC testing without their provider, but then expect their provider to interpret the results. Many of our colleagues experience this scenario on a weekly basis. While the existence of these services encourages patients to investigate their own disease vulnerabilities, and promotes engagement in their health, it also challenges the process of delivering sound medical advice. It is, therefore, critical to evaluate the quality and reliability of the genetic test results themselves in response to these queries. A recently published analysis of DTC services found that all phases of DNA testing and analysis suffered from quality control issues, which impacts both the accuracy and the interpretation of the results.[4] In addition, the limitations of the DTC genotyping panels mean that a "positive" test result may indicate a need for further consultation (though subject to the vagaries involved in variant classification described earlier), and that a "negative" result is rarely meaningful, given that most variants are not evaluated in these tests.

Despite the limited utility of DTC test results in most cases, the complexity of the processes and the hype generated by the press have undermined public awareness of the potential hazards. Consider the example illustrated by SNP variants reported in the methylenetetrahydrofolate reductase (*MTHFR*) gene. The MTHFR enzyme is responsible for converting the amino acid, cysteine, into another amino acid, methionine. When MTHFR's function is impaired by a variant in the *MTHFR* gene, the result is an elevation of homocysteine and a depletion of folate levels.[5] Here we describe a case that involves the younger of two brothers, aged 45 and 43, of northern European descent, who developed a deep venous thrombosis during a flight (designated as III-2 in Fig. 9). A few years earlier he had experienced leg swelling after a long automobile trip, but did not seek medical attention. His physician inquired about his family health history which revealed that his father (II-1) had died from a heart attack at the age of 52 and his paternal uncle had died from a pulmonary embolism at age 49 (II-2). Both were heavy drinkers and smokers, and suffered from depression. In addition, his paternal grandfather (I-2) also suffered from depression requiring treatment with lithium, and died of a heart attack in his 70s. At the time of the physician visit, the patient had lost contact with his father's siblings, and did not have any additional information about his paternal relatives. When encouraged by the physician to gather additional medical history, given the dangers of thrombosis and embolisms, the patient's mother re-contacted his father's family and learned that a paternal aunt had died from a pulmonary embolism in her 50s. In addition, the aunt's 24-year-old daughter (the patient's cousin, III-3) had developed pre-eclampsia, prompting her provider to test for three SNPs: Factor V Leiden, prothrombin G202010, and one in *MTHFR*. The first two variants are found in 1–5% of European (and some other) populations, and elevate the risk for developing deep venous thrombosis by 3- to 8-fold. Combined, they are multiplicative and can increase the risk for thrombosis 20-fold.[6] The *MTHFR* variant, however, does not have good evidence linking it to thrombosis. Despite this there is a great deal of conjectural and unsupported information reported

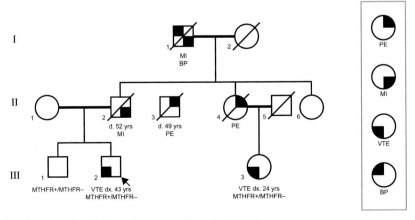

FIG. 9 Family health history for case involving the *MTHFR* gene.

in the lay press and online. The cousin was negative for both pro-thrombotic SNPs, but positive for the *MTHFR* SNP. Given these results, the two brothers with whom we started this story, were both tested for the 3 SNPs. Both were found to only have a variant in the *MTHFR* gene. Based entirely on these results, the younger brother declined to treat his thrombosis with blood thinners, the only proven efficacious therapy, and instead opted to initiate a dietary regimen intended to ameliorate the impaired enzymatic conversion to folate caused by the *MTHFR* variant. A risky and potentially fatal decision. Ultimately, after repeated appeals from his provider, he did concede and started blood thinners.

Three points are illustrated in this case:

- The value of a genetic test depends upon a cause and effect relationship between genotype and phenotype. Early *MTHFR*-variant studies assessing its relationship to thrombosis, heart attacks, and psychiatric disorders were carried out in small unselected populations. No association was found in later higher quality studies. The failure to replicate genetic association studies in different and diverse populations is common, largely due to factors such as poor enrollment criteria, poor criteria for assigning disease status, limited genetic diversity of the study population, and the failure to collect the presence/absence of specific environmental triggers in the study population.
- Risk for disease cannot be inferred for heterozygotes (one strand of DNA carries the variant) from homozygotes (both strands carry the variant) (Fig. 10). The original *MTHFR* studies reported an association between homozygotes and the presence of disease, not heterozygotes. All relatives in the family were heterozygotes for the *MTHFR* variant.
- Variant prevalence in the underlying population is critical for assessing its clinical utility. In this case, the association of a very common variant with a relatively infrequent clinical phenotype (thrombosis) suggests a very low penetrance and/or no true association at all. The *MTHFR*-variant in this family is commonly found in populations of European descent. About 16% of carry two copies and almost 50% carry one (thus it has been present in the human population for thousands of years). It is essential that studies match the genetic background of case and control subjects to avoid false positive associations.

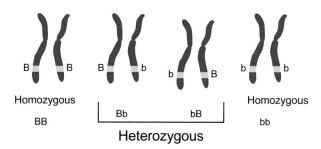

FIG. 10   Homozygous and heterozygous.

We will conclude this chapter with one additional demonstrative case, re-lated to *APOE*, a gene included on many DTC tests that has medically relevant insights. The apolipoprotein E (*APOE*) gene has three variants involving two different SNPs in its protein coding region. The ancestral variant, *APOE*e4*, carries neither of the two SNPs, and is associated with two highly prevalent diseases: Alzheimer's disease and hypercholesterolemia. While high cholesterol is imminently treatable, Alzheimer's unfortunately is currently neither prevent-able nor treatable, and its incidence is rising exponentially, as is its impact on society in terms of cost and quality of life. In addition, the impact is likely to get worse, as most of the world's population is aging, and the *APOE*e4* variant is common worldwide (~30%).

The human *APOE*e4* variant is identical to the only sequence of the *APOE* gene found in chimpanzees and explains its current high prevalence in many populations across the World. Data suggest that the APOE4 protein selectively improves survival when food supplies are limited and sporadic, which character-izes the human environment throughout most of our evolution. Today carriers with one copy of the *APOE*e4* variant (heterozygotes) are twice as likely to de-velop Alzheimer's disease by age 85 than those without it; and carriers with two copies (homozygotes) are eight-fold more likely to develop the disease, often at a younger age. However, the *APOE*e4* is a low penetrance variant associated with a common disease. Interestingly, the risk for developing Alzheimer's is further elevated among carriers who do not complete high school, though the nature of the relationship is unclear. Is this risk suggesting the importance of schooling and mental activity for later cognitive function? Or does this additional risk result from the impact of social determinants of health and lifestyle habits?

How should providers interpret a DTC genetic test that shows their patient carries one or two copies of the *APOE*e4* variant? This is a question that must be answered carefully, as illustrated by our second case. A 56-year-old man, who underwent DTC testing, discovered he was heterozygous for the *APOE*e4* variant, and therefore at slightly elevated risk of Alzheimer's disease (Fig. 11; IV-2). The man, whose father was confined to a nursing home after developing Alzheimer's in his late 70s (III-2), was extremely concerned. His paternal aunt (III-3) had also developed high cholesterol and Alzheimer's in her late 70s. He shared his DTC results with his relatives and found that one of his paternal cousins, through his affected aunt, also was heterozygous for *APOE*e4*. As the man conversed with other family members, he was reminded that his paternal grandmother (II-2), who had no serious prior health concerns, had died from a sudden heart attack at age 74; and that his father had suffered a mild heart at-tack, also at the age of 74, even though he was taking a statin for his cholesterol. His provider advised him to simply exercise, eat sensibly, and continue taking his prescribed statin. The provider also noted that there were no interventions currently available for preventing or reversing symptoms of Alzheimer's.

There is an underlying but relevant fallacy embedded in the case above: when interpreting familial disease patterns, disease prevalence needs to be taken in to

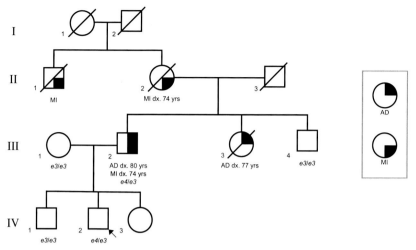

FIG. 11 Family health history for case involving *APOE* gene. *e4* refers to the ancestral *APOE*e4* variant, and *e3* refers to a variant resulting from a single SNP in its *APOE* gene sequence.

consideration. In this case, the man assumed that his father and paternal aunt had inherited an *APOE*e4* variant from his paternal grandmother. However, 40% of Alzheimer's cases are not due to the *e4* variant, and the vast majority of late in life heart attacks are not either. It is entirely conceivable that the *APOE*e4* variant, which is common in populations throughout the world, came from the patient's paternal grandfather, who was not afflicted by either Alzheimer's or heart disease, and who, of course, had never been genetically tested. Once again, a consideration of the patient's personal and family health history may reveal clinical information that refines both risk and prognosis concerning heart disease and other conditions that would not be further clarified by adding genetic testing. While DTC companies try to state the limitations of their tests on their websites, the concepts are so inexplicably foreign to those not trained in genomics that it is next to impossible to properly inform them using the currently employed methods.

In the future we may better understand the relevance of these types of genetic findings. For example, new studies have highlighted a role for SNPs in other genes to augment the risk for Alzheimer's disease in those with the *APOE*e4* variant. We will discuss these multi-SNP risk effects (also known as polygenic risk scores which were briefly described in Chapter 2) in more detail later in the book. These studies pose interesting implications for families, since they imply that there are other SNPs which individually and collectively contribute to heart disease risk, and whose effect can be further elevated by the additional presence of the relatively common *APOEe4* variant. One important aspect of these results is that the impact of the *APOE*e4* variant alone may not be enough to robustly increase disease risk, but the family's health history can serve as a reporter for the family's susceptibility to heart disease and Alzheimer's disease.

This raises questions that as of yet have no firm answers: How many variants exist in the human population that could lead to an elevated risk for a specific disease, health event, or condition? How feasible is it to employ predictive genetic testing to assess such risks? Finally, how variable are the effects of known pathogenic variants in different family contexts?

One additional comment about risk assessment, before we leave Alzheimer's disease, is the usefulness of testing for a disease with a late onset and no cure or preventative interventions. With the advent of risk-based testing in healthy individuals, as opposed to searching for a diagnosis in those who are ill, test results have been classified into 3 broad categories: not serious, serious without any available interventions, and serious with interventions available. Alzheimer's disease would be classified as serious without any available interventions, though compared to Huntington's disease, which would also fall into this bucket, it is far less serious due to its later onset. Each individual should prospectively opt into or out of receiving results for each category. The choices will differ not only between different individuals, but even between different members of the same family. Some may want to know everything about their risk regardless of whether they can intervene or not, some will only want to know if they can do something about it, and others may not want to know either. It is important to talk through these options and better understand what benefit the individual hopes to obtain from undergoing testing. In addition, they may want to talk with their family members about whether they would want to know his/her results. In some cases, relatives may not only not want to know about their own status, but also not want to know that something runs in their family. In my clinical career, I have had several cases where a parent underwent testing to help inform a child's risk, and the child chose not pursue testing even in the face of the parent's pathogenic results. In these situations, when you expect a relative to have one reaction and they have the opposite, it can be very unsettling. Considering potential challenges prior to testing can help, but it's hard to anticipate all of them. The best advice, given where we are now, is to proceed thoughtfully, investigate the limitations of the different test options, and communicate clearly with family members.

Lori A. Orlando is the lead author for this chapter. First person statements are from her point of view.

References

1. Manrai AK, Funke BH, Rehm HL, et al. Genetic misdiagnoses and the potential for health disparities. *N Engl J Med*. 2016;375(7):655–665. https://doi.org/10.1056/NEJMsa1507092.
2. Plon SE, Eccles DM, Easton D, et al. Sequence variant classification and reporting: recommendations for improving the interpretation of cancer susceptibility genetic test results. *Hum Mutat*. 2008;29(11):1282–1291. https://doi.org/10.1002/humu.20880.
3. Richards S, Aziz N, Bale S, et al. Standards and guidelines for the interpretation of sequence variants: a joint consensus recommendation of the American College of Medical Genetics and Genomics and the Association for Molecular Pathology. *Genet Med*. 2015;5(10):30.

4. Tandy-Connor S, Guiltinan J, Krempely K, et al. False-positive results released by direct-to-consumer genetic tests highlight the importance of clinical confirmation testing for appropriate patient care. *Genet Med.* 2018;20(12):1515–1521. https://doi.org/10.1038/gim.2018.38.
5. Liew SC, Gupta ED. Methylenetetrahydrofolate reductase (MTHFR) C677T polymorphism: epidemiology, metabolism and the associated diseases. *Eur J Med Genet.* 2015;58(1):1–10. https://doi.org/10.1016/j.ejmg.2014.10.004.
6. Kujovich JL. Factor V Leiden thrombophilia. *Genet Med.* 2011;13(1):1–16. https://doi.org/10.1097/GIM.0b013e3181faa0f2.

Chapter 4

Family-specific genetic variants: Principles, detection, and clinical interpretation

Brian H. Shirts, Vincent C. Henrich, and Lori A. Orlando

- The scope and scale human genetic variation.
- Population and family-specific variants can cause high risk and are responsible for most common autosomal recessive diseases.
- Family-specific variants can cause high risk and are responsible for most autosomal dominant disease.
- Classifying rare and family-specific variants requires application of specific criteria.
- Variants are most often classified on a 5-level scale from pathogenic to benign.
- Family-specific variants are often classified as variants of uncertain significance (VUS).
- Several different types of evidence are used to understand family-specific variants and classify VUS.
- Families are among the most efficient ways to ascertain clinical data about VUS.

Introduction—The scope and scale human genetic variation

Genome sequencing in many species has revealed that of genetic variation has been present since before the dawn of *Homo sapiens* as a species; many uniquely human variants have been in human populations since before humans began to explore beyond the African continent. As discussed in the previous chapter, these ancient genetic variants result in common variation that can be seen throughout current human populations. These, along with other common variants that initially arose in the human population thousands of years ago, have been widely studied through genome wide association studies. Genome wide association studies (GWAS) as used to identify genetic disease risk were discussed briefly in the previous chapter. Although there have been several

Managing Health in the Genomic Era. https://doi.org/10.1016/B978-0-12-816015-2.00004-3

moderate risk variants discovered by GWAS, the vast majority of findings have been of low-risk variants. Evidently, and as expected, natural selection has prevented high-risk and disease-causing variants from being common in the general population. It is obvious that variants that are incompatible with reproduction are selected from the population each generation. Variants that cause a high risk for severe adult onset genetic diseases have also been subject to high selective pressure. This chapter will discuss variants that are much more recent, much less frequent, and which may predispose someone to higher disease risk than variants identified by GWAS studies that are used to calculate polygenic risk scores.

Mutation is the ultimate source of all variation in the human genome. When a mutational event occurs, it creates a change in the DNA sequence, which is copied as a cell replicates. If the mutation is passed on to the next generation it is referred to as a genetic change that with a frequency in the population may be referred to as a genetic variant. Genetic variants that are present in over 1% of the population are referred to as genetic polymorphisms. Historically genetic variants that caused disease were called "mutations"; however, this term may not be accurate and has incorrect negative connotations for some (or perhaps connotations of super powers for Marvel comics fans).[1] As our understanding of the broad spectrum of disease risk influenced by human genetic variation has grown, the term "mutation" has been replaced by "pathogenic variant" as the preferred term for disease causing variants.[1]

The human mutation rate has been estimated to be between 1×10^{-8} and 2×10^{-8} single nucleotide changes per base pair per live birth.[2,3] The frequency of new indels (insertions and deletions) and larger genetic changes per live birth is about an order of magnitude lower, even though these genetic changes may be more likely to have an effect on human health. Keeping in mind that over 3 billion base pairs comprise the haplo-genome of each sperm and egg cell, this translates to between 35 and 75 new single nucleotide changes and between 10 and 20 other genetic changes per live birth (Fig. 1).

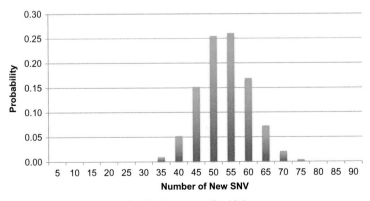

FIG. 1 Number of new single nucleotide changes per live birth.

As the human population has grown over the last several thousand years many of these new mutations have been passed on to offspring in subsequent generations, becoming genetic variants. Genetic variants have been accumulating over time as the human population grows. With dozens of single nucleotide changes per birth and 9.6 billion possible single nucleotide changes in a haploid genome of 3.2 billion base pairs, as a society we should expect to find every single nucleotide change that is compatible with life as we continue to sequence more and more patients. Indeed, every single nucleotide change that is compatible with reproduction is almost certainly carried in a family alive today. The total number of genetic variants in a population is related to size of the population, until the population is large enough that all variants are present. All single nucleotide changes compatible with life were probably present in the human population at somewhere around a population size of 1 billion humans (see Fig. 2). (Because deletions are variable in size and because insertions can be any size or sequence, there can theoretically be an infinite number of indels (insertions and deletions).)

As we have already described, each inherited genetic variant in the population initially arose in a single founder individual. Because the human population has grown relatively rapidly over the last several centuries, most genetic variants are relatively recent in origin. It follows that the less frequent a variant the more likely it is to have occurred relatively recently. Many of the extremely rare variants we see now in clinical testing first occurred in ancestors that were alive within the last several hundred years. We will call these relatively recent genetic variants "family-specific variants".

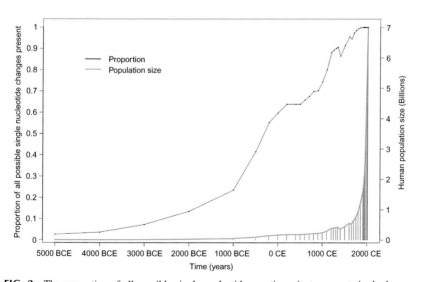

FIG. 2 The proportion of all possible single nucleotide genetic variants presents in the human population. This proportion is related to the population size in relation to the size of the genome. Over 99% of all single nucleotide substitutions compatible with life have been present in at least one living human for about 200 years.

All of us have many family-specific variants, some that we share with a handful of close relatives, some that we share with thousands of distant cousins, and everything in between. Even though each of us has many of these variants, the set of family-specific variants that anyone has is as unique as their family tree. There is not a lot known about most of these specific variants. A few variants found their way into the hands of genetic researchers in time to be included in seminal publications. A larger number have listings in public databases, such as ClinVar (https://www.ncbi.nlm.nih.gov/clinvar/). But most family-specific variants have yet to be listed or identified in any database. Most of these family-specific variants, and most variants in the human genome, are benign and have no discernible effect on any disease risk or any human trait. However, some population and family-specific variants can cause disease or increased disease risk.

Population and family-specific variants can cause high risk and are responsible for most common autosomal recessive diseases

There are many examples of medically important variants that are specific to known ethnic populations and are relevant to healthcare in adults (such as those listed in Chapter 3, Table 1). The best known and studied variants that cause recessive disease are specific to European populations. These include: *CFTR* delta F508, a deletion of one phenylalanine amino acid, which is by far the most common variant that causes cystic fibrosis. *HEXA* 1278insTATC, the four base pair insertion present in one in every 27 people with Ashkenazi Jewish ancestors, causes Tay-Sachs disease. *GBA* N370S, a single amino acid change, is the most common cause of Gaucher's disease, also in the Ashkenazi Jewish population. There are other well studied variants specific to other racial and ethnic groups: The *HBB* gene variant known as Glu6Val (E6V) or HbS described in Chapter 3, which is carried by 1 in 11 Africans and causes recessive sickle cell anemia. Several alpha-thalassemia causing deletions are common in different populations. The single alpha-globin deletion $-\alpha^{3.7}$, which is more likely to cause mild microcytic anemia in adults than any symptoms, is very common in Northern and equatorial Africa. Two gene alpha globin deletions known as SEA, THAI, and FIL, which cause microcytic anemia in the heterozygous state and Hb Bart hydrops fetalis in the homozygous state, are relatively common in Southeast Asian populations.

One of the most common medically relevant population specific variants is the *HFE* C262Y variant, which is carried by 1 in 8 Europeans and homozygous in 1 in 200 Europeans, and is associated with a recessive predisposition for hemochromatosis. The *HFE* gene is involved in transferrin regulation. The C262Y variant increases the absorption and storage of iron potentially leading to iron overload and liver failure. The variant is population specific, as it is virtually absent in Asian and African populations. One reason why this variant may have become so common in the European population is that only a small percentage of people homozygous for this variant have negative effects. This is an example of reduced penetrance of a risk allele. Iron overload symptoms due to *HFE* C262Y are more common in men with other diet and lifestyle factors such as iron intake and alcohol consumption playing a role. Therapeutic phlebotomy can treat and prevent iron overload, so developing symptoms is extremely unusual in regular blood donors. Although this is one of the most common and well known population specific variants, like other population specific variants, there are ongoing debates about the relative utility of genetic screening in the at risk population.[4–6]

Populations with other well-known specific mutations may have had some degree of cultural or geographic isolation, and can sometimes trace these variants to a group of historical founders. We have already noted two variants known to be Ashkenazi Jewish founder variants. Others founder variants have been described in Finnish, French Canadians, Icelandic, Venezuelans, Brazilians, as well as in other groups (see Chapter 5). The known population specific variants tend to be more common in populations where research has been well accepted and in countries with research budgets and advanced health care infrastructure to support genetic testing. The knowledge of founder variants in these populations is undoubtedly because that is where research has been done. As genetic sequencing has become more common around the world, there have been an increasing number of reports about other founder variants in many populations.

Population-specific founder variants in well-described populations receive most attention in textbooks, including this book. Although some examples of recessive disease have adult onset symptoms, like hereditary hemochromatosis, or may be detectable but not symptomatic in the heterozygous state, like alpha-thalassemia, most recessive genetic diseases have pediatric onset. The disease status of siblings and children can be very important to note in family health histories. A physician who is aware of relatively common recessive diseases and associated traits will be able to counsel patients about their own risk and know when to pursue further genetic testing. For example, it is common to see physicians ordering complex anemia workups when an informed family historian might quickly deduce situations where the most likely cause for microcytic anemia is an alpha globin deletion rather than iron deficiency. This type of family history combined with racial and ethnic awareness will take continuous effort. There are certainly also clinically important founder variants in less well-studied populations, which will become apparent as more individuals from diverse populations are sequenced.

Family-specific variants can cause high risk and are responsible for most autosomal dominant disease

In contrast with autosomal recessive traits, well-described population specific variants cause a small proportion of autosomal dominant traits. For example, the three most common Ashkenazi Jewish *BRCA1* and *BRCA2* variants are present in

2% of the population and responsible for 59% of the hereditary breast and ovarian cancer in the Ashkenazi Jewish population, but these three variants account for a very small portion of all hereditary breast and ovarian cancer in the general population.[7-9] There are many autosomal dominant genetic diseases that are incompatible with reproduction or that cause symptoms so severe that reproduction is unlikely. These are most likely to be seen as de novo or sporadic mutations in isolated individuals. It has become clear that many highly deleterious mutations have no clinically observable phenotype because their effects are disruptive to basic cell metabolism or early fetal development. Therefore, they never survive.

Variants that lead to the autosomal dominant diseases or predisposition to diseases most likely to be treated by internists are often caused by variants that might be best described as lineage- or family-specific variants. When not everyone with a risk variant develops an associated disease, this is referred to as incomplete or reduced penetrance (see Chapter 5). The time of onset may also be correlated with age. Because adult onset dominantly inherited disease predisposition variants cause increased morbidity and mortality, there has been evolutionary pressure against these variants. However, with recent explosive population growth there may be hundreds of thousands of variants that cause autosomal dominant disease that have only been around for a few generations.

These family-specific variants cause a wide variety of adolescent and adult-onset diseases. One example is hypertrophic cardiomyopathy. A classic case is that of a healthy 19-year-old male who died suddenly while playing soccer with friends. Autopsy revealed a hypertrophied left ventricle. A family history was collected, and which showed a family history of later onset cardiac disease in several relatives but was unremarkable for early cardiac disease, with the possible exception of an early unexplained death in a maternal great-aunt (Fig. 3). Genetic testing from autopsy tissue identified a pathogenic variant in *MYH7* which had been reported previously in two other families in public databases. *MYH7* is one of over two dozen genes associated with hypertrophic or dilated

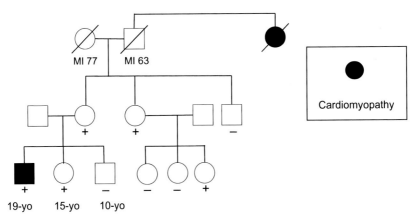

FIG. 3 Pedigree of example family with family-specific variant in *MYH7*.

cardiomyopathy. Subsequent testing showed that the mother had the same *MYH7* variant, and had a normal ECG with borderline ventricular hypertrophy on echocardiogram. As other family members were told about the genetic findings additional maternal relatives were tested. It was thought to be likely that the maternal great-aunt had the *MYH7* pathogenic variant, and several other relatives with the variant had cardiology consultations about risk of sudden death. Beyond the maternal great aunt, there were no additional distant relatives that had suspicious reports of sudden death.

This example family was fortunate that the variant identified had been reported in other families and already classified as pathogenic. The evidence from the 2 other families in public databases helped clarify the classification and implications of the variant for this family. If the variant had not been reported in public databases, it is likely that this variant would have initially been classified as a variant of uncertain significance, a term that will be described later. In addition, it is likely that the two families from the literature are distant cousins of the family in the clinical case. Although it is possible that this variant is the result of multiple unique mutational events, given the low human mutation rate at each unique position, it is much more likely that there was only one initial mutational event and that all three families inherited the variant from the same ancestor. This is not a population-specific variant, rather it is a family-specific variant.

What is the expected distribution of family-specific variants in the population and what variants are you likely to see?

A comparison of two populations where there has been comprehensive evaluation of all variants for one disease, Lynch syndrome, illustrates how high-risk variants for adult onset disease are distributed in different populations. Lynch syndrome, also known as hereditary nonpolyposis colorectal cancer (HNPCC), is caused by pathogenic variants that disrupt the *MLH1*, *MSH2*, *MSH6*, and *PMS2* genes, which are involved in DNA mismatch repair. It is associated with over 50% lifetime risk of colorectal and endometrial cancer as well as elevated risk of other, less common, cancers and will be described in more detail in Chapter 5.

A nationwide analysis of Lynch syndrome of all colorectal tumor in Iceland identified all recurrent mutations in the 4 mismatch repair genes known to cause Lynch Syndrome.[10] The most common variant in Iceland, PMS2 c.736_741del6ins11 (see box on next page), has also been reported in families in Sweden, Britain and the United States, and has been dated to have arisen around 1625 years ago.[11] Four other Lynch syndrome mutations were observed in single families that could be traced back to a common ancestor that lived in the 1700 to 1800s. There were several other variants seen in one family each. The Icelandic population is ideal for this type of comprehensive study of a well-described autosomal dominant disease because the population of 300,000 has been relatively stable with little immigration. Thus, the genetic variation within the population was not introduced by an influx of migrants and the population can be viewed as one in which the variants that currently exist there arose almost solely in an individual founder. The frequency of a variant in the population is correlated with the age of the variant and therefore, newer variants are less frequent. So, the fact that most Lynch syndrome variants are unique to small families illustrates that most highly penetrant dominant disease variants are recent in evolutionary history.

Understanding variant annotation

Initially researchers that identified new variants named the variants they discovered. To avoid confusion, the Human Genome Variant Society has created guidelines for naming and reporting variants. Here is a quick overview of common features you may see in a variant report:

Gene name: *BRCA1*—gene names are all caps and usually in italics. Corresponding protein names are usually not italicized.

Nucleotide change: c.5324T>A—This indicates the nucleotide position in the gene, usually counting starts from the first coding nucleotide in the gene. T>A indicates a thymine was changed to an alanine in the DNA transcript for this gene. "del" or "dup" indicates that the nucleotides numbered were deleted or duplicated. Sometimes there is a "−" or "+" before the number, which usually indicates a splice site change that is at a nucleotide just before the beginning of an exon or just after the end of an exon.

Protein change: p.Met1775Lys. This indicates the protein change that is caused by the nucleotide change. The reference amino acid is before the numbers (Met) the resulting amino acid is after the numbers (Lys). Three letter amino acid cods are more common, but sometimes single letter codes are used (p.M1775K is the same as p.Met1775Lys). If change leads to a stop codon this is indicated by Ter, *, or X. If a deletion or duplication leads to a translational frameshift the letters "fs" indicate this, and the length of sequence before the new stop is sometimes also listed.

Larger deletions, duplication can also be listed. These annotations are usually self-explanatory. There also guidelines for annotating inversions, translocations, and truly complex genomic changes. For a more complete understanding of variant nomenclature please look up the Human Genome Variant Society website (www.hgvs.org).

Studies in larger populations have yielded similar findings. Data from two sequential comprehensive studies of colorectal cancer in central Ohio revealed the genetic architecture of hereditary colorectal cancer in this region.[12] These studies found 24 variants in the *MSH2*, *MSH6*, *PMS2*, and *MLH1* genes that were identified in multiple individuals. Statistical estimates were then used to predict the number of Lynch syndrome variants in the population of about 5 million people in central Ohio that was covered by the study. The resulting estimate was that 243 family-specific variants in central Ohio cause Lynch syndrome (Fig. 4). These variants are likely to be seen in an estimated 1508 families (when a family is defined as three generations of individuals).

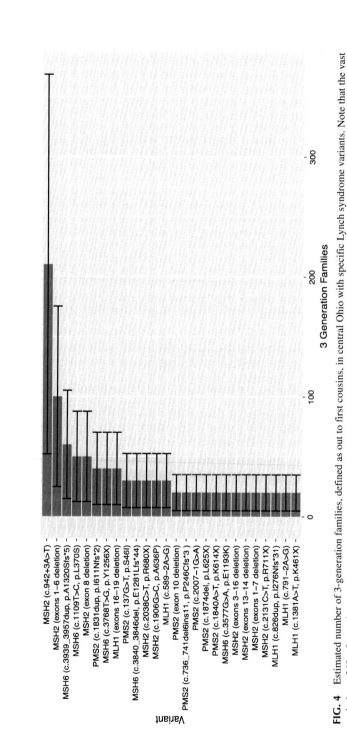

FIG. 4 Estimated number of 3-generation families, defined as out to first cousins, in central Ohio with specific Lynch syndrome variants. Note that the vast majority, > 260, of mutations we expect will be present in a small number of families and are not plotted. Error bars given here are the 95% confidence intervals of the estimates.

Analysis of haplotypes, or sets of variants that are inherited together on the same DNA strand, revealed that the most frequently observed variant, $MSH2$:c.942+3A>T, had at least three founders. This is a variant that is a known mutational hotspot in the $MSH2$ gene, which accounts for its reoccurrence in a single population. Most other variants appeared to have a single founder. Models indicate that in $MSH2$ the second most frequent variant is a large deletion of exons 1 through 6 and is between 425 and 625 years old.[13] Genealogy identified a common ancestor of individuals with $MSH6$:p.L370S, the fourth most common Lynch syndrome variant observed in this region who was born in 1844.

Findings in Ohio are similar to those from Iceland and to those from other studies in different populations. Although there are a few mutations that arose more than 500 years ago, most medically relevant Lynch syndrome variants might be traced to ancestors that were alive within the last few hundred years. The other revealing finding is that the overlap of high-risk genetic variants between the Ohio and Iceland populations is almost non-existent. This should not be surprising since variants arose within the last few hundred years. It is similar to what is seen by studies of hereditary breast and ovarian cancer in diverse populations.[7] These studies for colorectal cancer risk genes reveal the pattern of population structure of medically important genetic variation in adults. Different populations with distinct ancestral histories are expected to have different sets of medically important genetic variants for dominant traits. Other dominant adult onset single gene disease traits are also caused by variants that have arisen in the last few hundred years and are similarly likely to be caused by family-specific variants. This means that although two physicians who work in different states may see patients with the same genetic diseases, the specific variants responsible for those diseases are likely to be different.

This also means that a physician is likely to encounter many more genetic variants than the small subset of model variants discussed during medical school and residency training. In clinical practice, with regard to dominant genetic risk, you are almost as likely to treat adult patients that have extremely rare or never-before-seen genetic variants as you are to treat patients that have the common variants you have heard about in medical school. The practical implication is that when genetic risk is suspected, ordering a test that targets a specific variant is less likely to identify the cause than a test that completely sequences all genes associated with the disease. If there is a risk variant, it is most likely to be one that you have never seen or heard about before, even if it is in the gene that you expect. Even in individuals with Ashkenazi Jewish ancestry and strong family history of breast cancer, targeted testing for only the three most common mutations has substantially lower sensitivity than the recommended panel sequencing for all mutations in $BRCA1$, $BRCA1$, $PALB2$, and other breast and ovarian cancer risk genes. When there is already a known pathogenic variant in the family, targeted testing may be indicated. However, without a family history of a specific, known genetic variant, larger panel testing of multiple genes associated with the family history of disease may be indicated. For very unusual presentations of rare familial disease exome or genome testing may even be indicated. The distinction between general genetic testing and more targeted testing is critically important for diagnostic purposes.

Appropriate genetic test ordering can be complicated. Because of large cost differences in genetic tests and complex insurance rules, it is often beneficial to ask someone experienced in genetic testing, a genetic counselor, or your hospital or reference laboratory for assistance with ordering a genetic test. Asking for help can help find the correct result, save time, and save money as discussed in more detail in Chapter 8.

Classifying rare variants requires application of specific criteria

The clinical corollary of the fact that many rare missense variants are in the germline of several living humans today is that when clinical sequencing is performed, there is a good chance it will uncover a variant that has never been identified previously or that has not been well classified. Only a small portion of possible variants in cancer risk genes has been reported and of the reported variants less than 5% have meaningful clinical interpretations.[14,15] Even though pathogenic variants for dominant diseases are uniformly rare and often family specific, the corollary does not hold. In reality, not all family-specific variants are associated with disease. The provider would do well to remember that most family-specific variants have no meaningful effect on health.[16,17]

Variants are most often classified on a 5-level scale from pathogenic to benign

For genes known to cause a disease, variants can either cause the disease or not, pathogenic or benign. However, when a variant is seen for the first time or has not been well studied, it is not always clear if the variant causes disease. Hence, variants are classified as pathogenic, likely-pathogenic, variant of uncertain significance, likely-benign, and benign (see Chapter 3). The middle three classifications do not denote separate biological phenomena, but rather levels of evidence. A report with a VUS does _not_ suggest intermediate risk, rather it indicates low certainty based on little or conflicting information about the variant (see Table 1).

Categories for data that can be used to classify variants are: population data, computational and predictive data, functional data, family segregation data, de novo data, allelic data, database information, and other (see below for a more complete discussion of these types of information). Several categories of data and examples of how these are used are described later in this chapter. There remains flexibility in how these rules are applied, which can result in differences in classification between clinical laboratory reports.[18,19] Because clinical

TABLE 1 Variant classifications, clinical actionability, and probability of pathogenicity associated with classification.

Classification of variant	Clinical actionability	Probability that the variant has the same effect as well characterized pathogenic variants in the same gene
Pathogenic	• May change prevention and treatment plan	>99%
Likely-pathogenic	• At-risk relatives should be tested	95–99%
Variant of uncertain significance	• **Should not alter medical care** • Consider revisiting classification in 1–2 years • Consider family studies to build evidence about the specific variant	5–95%
Likely-benign	• Should not alter medical care	0.1–5%
Benign		<0.1%

Bold type indicates the recommendation most often missed.

judgment and expertise are necessary to apply these rules, it is recommended that molecular pathologists and medical geneticists experienced in variant assessment apply these rules to interpret clinical variants.

Family-specific variants are often classified as variants of uncertain significance (VUS)

There is a linear relationship between length of DNA evaluated and number of rare variants of uncertain significance (VUS) reported in clinical sequencing reports (Fig. 5). If a family-specific variant is found in a gene associated with disease, and the biological relevance is not immediately clear, as is the case for loss of function nonsense or frameshift mutations, the variant is almost always initially classified as a VUS after applying variant classification guidelines.

The common reporting of VUS is a relatively recent development in medical genetics.[20–24] The term VUS was popularized by Myriad Genetics when it started doing commercial *BRCA1* testing. The company quickly realized that there were many patients with unique variants for which they did not have any information. The VUS term acted as a legal place holder, so that the testing company could return a result that was not clinically meaningful, but was also not demonstrably negative. Although this classification was not created to improve the medical management, it has stuck, and a large body of literature and standards of medical practice have arisen around this designation.

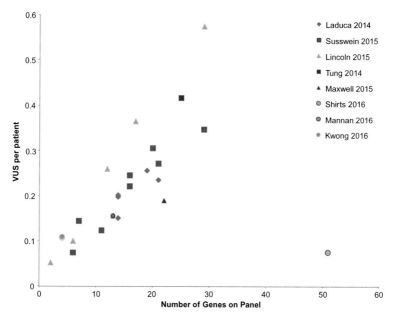

FIG. 5   Average number of variants of uncertain significance relative to number of genes sequence. The graph presents the linear relationship between the average number of variants of uncertain significance reported per patient tested for hereditary cancer risk, and the number of genes sequenced at most commercial laboratories. Data were collected from published manuscripts by clinical laboratories performing sequence interpretation.

The American College of Medical Genetics recommends that if a VUS is identified, this information <u>not</u> be used in medical management decisions.[25] Instead medical decision making should rely on personal presentation and family history. This means that VUS should be treated almost exactly like benign variants, with the exception that patients with a VUS may be encouraged to follow up by periodically (every 1–2 years) checking in with their providers to see if the classification has changed. Patients with VUS may also be encouraged to participate in activities such as family studies or RNA analysis to find out more about their VUS. Table 1 gives a summary of general medical actionability for different classes of variants.

In variant reclassification studies, the majority of rare, or family-specific, variants in genes that have been associated with disease turn out not to change disease risk, and are eventually classified as benign.[17,23] The fraction of VUS that are eventually reclassified as benign varies by gene, with about 80% of VUS in *BRCA1* and *BRCA2* eventually classified as benign and about 60% of VUS in *MLH1* and *MSH2* classified as benign. The fact that most variants in genes that are known to be associated with disease are initially classified as VUS and eventually turn out to be benign highlights the importance of accurate communication about the medical implications of VUS so that patients and other healthcare providers base their treatment on **family health history**, rather than the presence of the VUS. **Variants of uncertain significance should not alter medical care**.

Communicating about variants of uncertain significance in clinical care

Communicating uncertainty about medical diagnosis is difficult. This is true for all aspects of medicine, so it should not be a surprise that communicating uncertainty about genetic findings is challenging. VUS results are particularly difficult for patients to interpret and recall correctly.[24] This may be because providers express uncertainty about what to do with a VUS and do not do a good job of explaining that VUS should not alter medical care. On the other hand, it may be because patients are prone to expect that genetic testing results in medically actionable steps, and have a difficult time understanding that this is not always the case. Misunderstanding of VUS may arise because of difficulty with probabilistic explanations, challenges relating to intrinsic ambiguity in the information, and discordant provider explanations about the clinical significance of VUS.[26] This lack of clarity may cause distress or anxiety for some patients.[27–29] Providers and patients may incorrectly act on VUS information because they think that VUS are clinically actionable. This misunderstanding and miscommunication has led to unnecessary surgery and inappropriately aggressive screening.[30–32] Needless to say, there have been lawsuits relating to inappropriate management of VUS.[33,34] The best way to address patient distress related to a family-specific VUS and to avoid inappropriate care is to understand recommendations and to be comfortable by not acting on uncertain information.

Understanding evidence behind variants and reclassifying VUS

For a family that has a unique VUS, correct classification is an imperative as VUS classification can improve medical outcomes, allow insurance payments, prevent unnecessary treatments, and reduce stress. Patients may choose to ask their providers if their VUS have been reclassified as there is more information gathered on variants every year. The time frame for reclassifying a VUS, however, is difficult to predict. One survey reviewing clinical reports from 4 years of testing found that 11% of observed VUS had been reclassified at some point, however many of the reports had only been issued 1 or 2 years earlier.[35] Another study found 23% of patients followed for at least 8 years still had a VUS result, suggesting that 77% of VUS had been reclassified over the 8 year timeframe.[32] A retrospective study spanning 17 years found that of 56% VUS were reclassified with a median reclassification time of 39 months for the subset of VUS that were reclassified.[31] Overall it seems that about 10% of VUS are classified per year.

One important exception to this slow classification rate is one study where patients proactively became involved in efforts to gather information from their own families. Over 50% of variants were reclassified in less than 1 year.[36] This study highlights one example of how active patient and provider efforts may speed reclassification efforts. Providers can contact laboratories to ask about details of variant reclassification. ClinVar[14] publishes lists of variants, with interpretations issued by different laboratories. If there is a discrepancy between laboratories, communication between those laboratories is likely to lead to VUS resolution and may lead to classification as the laboratories communicate and share data about variants.[19] Often it is the active query of a dedicated genetic counselor or provider that prompts a laboratory to seek additional information from about a variant.

Most laboratories offer family studies for VUS reclassification to selected families.[37] As noted above, active participation of patients in family studies for VUS reclassification led to reclassification of 60% of variants for participants that had been actively engaged in variant classification for at least a year.[36] Family studies are usually not sufficient to classify variants on their own, but can add to public information available from other sources.

Several different types of evidence are used to understand family-specific variants and classify VUS

There are several lines of evidence that are used to classify new genetic variants in clinically tested genes. In this section we will discuss several examples of data that can be used to classify a VUS that has only been observed once or a few times. Family-specific variants are a situation where randomized controlled trials are impossible, so it becomes necessary to use other sources of evidence in medical decision-making.[38] These types of information are detailed in ACMG-AMP recommendations for variant interpretation[25] and many subsequent papers comment on the proposed strategy.[18,39–43] The most detailed and comprehensive refinements are being implemented by the ClinGen (https://clinicalgenome.org/) gene specific working groups.[44] ClinGen is consortium of NIH funded and volunteer experts dedicated to generating consensus about high-quality clinical variant interpretation.[45] An overview of several types of evidence with examples of how they can be applied to specific variants are given here:

In-silico data

There are dozens of computational programs that have been created to calculate and model variant function.[46,47] These produce what is known as "in-silico" variant function predictions. Some use matrixes of protein changes that attempt to characterize the extent that amino-acid substitutions are predicted to alter protein structure. Others quantify evolutionary conservation. Splice prediction algorithms compare test sequences defined by machine learning with canonical splice motifs in order to assess the possibility that exon-intron splicing is disrupted by a specific variant. Some programs use combinations of different computational strategies. Computational algorithms are tested against a training set of variants and, ideally, validated against a separate test set. With this type of controlled data set currently reported algorithms boast greater than 80% sensitivity and specificity. Unfortunately, many training and test sets are highly selected, and although easy to use, in-silico algorithms may have low positive predictive value if used in a screening setting.[48]

Because of these limitations, computational tools are usually applied cautiously. In one recent example, Thompson and colleagues validated a combination of the MAPP and Polyphen2 algorithms for classification of variants in *MLH1*, *MSH2*, and *MSH6* genes associated with Lynch syndrome. While their own validation data suggested that several variants had probability of pathogenicity greater than 99%, for their final Bayesian classification calculations this group chose to truncate prior probabilities at 90% on the high end and at 10% on the low end. This practice of limiting the strength of information from computational models is standard for variant classification researchers. This avoids some of the concerns about overfitting and about algorithm validation datasets not being representative of real-life experience. ACMG-AMP standards consider most computational predictions to be supporting evidence at best.

Despite concerns about the positive and negative predictive value of in-silico data, these are widely if not universally, used in variant classification. The main reason is that these are incredibly easy to generate and access. Many researchers that develop computer prediction algorithms will generate functional scores for every possible single nucleotide change in the genome to go along with publication of their methods. This type of practice makes computational predictions for a rare variant extremely easy to obtain. When epidemiologic data on a variant is non-existent, computational data may be all there is to go on.

Functional studies

Functional studies are experiments that test a variant's effect on a gene in an experimental system. There are countless types of functional studies. These can be done in test tubes, yeast cells, or mammalian or human cell culture systems. Many initial functional studies were painstakingly designed to test individual variants that had been identified in research families. Increasingly, with the help of next generation sequencing and robotic laboratory systems, functional assays are being designed for high throughput analysis.[49] The quality of evidence derived from a functional assay depends on how well the test system matches the clinically important aspects of the variant's molecular biology, and how extensive the clinical validation has been to match assay interpretation to clinical outcomes.

ACMG standards allow using "well-established in-vitro or in-vivo functional studies" as strong evidence in favor of both benign and pathogenic variant classification. These standards give minimal guidance about what constitutes a "well-established in-vitro or in-vivo" functional study. There are currently no standards for reporting or validating functional assays for variant classification, so accuracy in application to variant classification is dependent on the user's knowledge of the functional test system. Regardless of the strength of the functional assay, this evidence is considered stronger than computational data, but generally not considered sufficient to independently classify a variant.

One situation where functional studies have high potential to be revealing is when a VUS is predicted to alter gene splicing. As described in the previous chapter, when DNA for a protein is transcribed to RNA, the RNA signal from a long DNA strand must be spliced, often at multiple sites, to generate the shorter mRNA that serves as a template for the translation of the protein. DNA variants that alter splicing efficiency are likely to be pathogenic by resulting in the translation of an incomplete protein. Functional analysis of splicing in patient derived RNA using RT-PCR or next-generation sequencing methods such as RNA CaptureSeq can show what proportion of a transcript has alternate splicing.[50,51] Not every intronic change affects splicing; if a variant that is predicted to change splicing does not cause alternate splicing it may ultimately be classified as benign. On the other hand, functional evidence of alternate splicing from RNA supports classification of a variant as pathogenic.

Families are among the most efficient ways to ascertain clinical data about rare VUS[52,53]

Family ascertainment is orders of magnitude more efficient than case-control studies for classification of rare variants.[54] First-degree relatives have a 50% chance of having the rare variant, which may be incomprehensibly higher than probability of finding a one-in-a-million variant randomly sampling from the population. Because of this being able to genotype a few relatives to help classify a family-specific variant is like winning the lottery. The relative utility of family-based approaches is illustrated by comparing two papers with very different approaches. An evaluation of 26,670 unique variants of uncertain significance variants reported between 2006 and 2018 that used population-based strategies was able to classify 7.7% of these unique variants in the study period.[17] In contrast, a smaller patient-driven study that tested a family-based approach in 92 families between 2016 and 2018 was able to classify 61% of variants of uncertain significance.[36] Although, it is not clear if the family-based approach can be scaled, it is certainly something that providers should advocate for.

There are several types of information to understand rare variants that can be gathered from families. These include cosegregation analysis, information showing a variant is de novo, haplotype phasing, imprinting information, and simply providing additional individuals with the genetic variant to evaluate genotype-phenotype correlation. Below we provide a complete explanation of cosegregation analysis with several examples, and brief examples of the other types of family information that can be used to classify rare variants.

Cosegregation analysis

Cosegregation analysis assess the probability that a variant and a trait travel together more often than expected by chance.[55] Practically speaking, the analysis provides a quantitative measure about the level of confidence that a specific variant is truly associated with the incidence of disease within a family. There are simplified methods of statistical analysis, such as counting meiosis,[56] which work well for highly penetrant traits. For later onset traits or disease that do not have complete penetrance, more sophisticated cosegretation tools far outperform.[57] Cosegregation analysis is accurate, if done well, but may not give enough evidence to classify a variant if only a small number of family members are available. For example, if two sisters who both have breast cancer and who both tested for *BRCA1*, happen to have the same variant, one might initially consider this very strong evidence that the variant is related to their breast cancer (Fig. 6). However, if the variant happens to be benign (and 70% of *BRCA1* VUS are eventually classified as benign) the probability that the two sisters share that benign variant by chance is 50%. This is a very small amount of evidence, and few physicians would suggest surgery if they thought there was a 50% chance that it would be unnecessary. So, a simple analysis would conclude that the variant is about twice as likely to be pathogenic. Although this is useful evidence, additional information is needed to reclassify a VUS and justify changing clinical management. Testing additional relatives may produce more evidence one way or the other, as cosegregation analysis can yield evidence that a variant is benign as well. All relatives at risk of having the variant regardless of disease status should be tested. Testing all close relatives that might have the same variant prevents bias that may skew understanding of the variant's function.[58] If only affected relatives are tested, with no unaffected relatives being tested, then many variants will eventually be classified as pathogenic simply because unaffected carriers that might prove the variant is benign have been ignored.

Another misconception with cosegregation analysis is that identification of the variant of interest in an unaffected relative indicates "Nonsegregation with disease," which is defined by ACMG-AMP criteria as strong evidence against

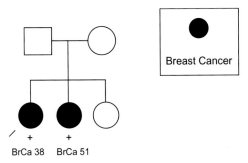

FIG. 6 Simple pedigree as an example of cosegregation.

pathogenicity. For genetic traits with complete or near-complete penetrance, where every individual with the genetic defect has symptoms of the disease, this may be true, but for traits with incomplete or reduced penetrance this is an over-simplification. One prominent example is cardiomyopathy. *MYH7, MYBPC3, TNNT2, TNNT3,* and many other genes cause hypertrophic cardiomyopathy; *TTN, LMNA, MYH7, MYH6, SCN5A, MYBPC3,* and many other genes cause dilated cardiomyopathy. There is large phenotypic variability with these syndromes, with some overlap in presentation between the syndrome. Many individuals with established pathogenic variants have mild or absent clinical features, even after in-depth cardiology workups. Because of incomplete penetrance many clearly disease-associated variants could be classified as benign based on rule-of-thumb co-segregation analysis. Cardiomyopathy expert panels have suggested not using family co-segregation analysis to classify variants in these genes. Quantitative cosegregation analysis is possible in this situation, but accurate penetrance models and appropriate statistical assumptions are necessary to get the right answer. These quantitative models are typically not accessible to primary care providers.

Examples of variant data that can be obtained from families

Because family data is useful, and is the type of data that is most readily available to most medical providers, I have included several examples describing how information on lineage specific variants can be gleaned from individual families. All example pedigrees and scenarios are presented to illustrate principles and have been modified from real families that I have seen. Because the pedigrees have been modified, no specific variants are described. These examples will give an sense of what types of results come from the statistical cosegregation analysis that evaluates if a variant travels with disease more often than expected by chance, as well as illustration a few other specific situations when understanding how a variant travels through a family can illuminate the clinical effect, or lack of clinical effect, of a family-specific genetic variant.

Examples of cosegregation analysis

1. *MSH2* variant with strongly suggestive cosegregation (Fig. 7).
   This family is not an unusual Lynch syndrome family. There are many relatives with colorectal or endometrial cancer, some of whom have very early onset, and some of whom do not (see additional discussion of Lynch syndrome in Chapter 5). The fact that the cousin of the proband has early onset colon cancer led to that individual being genotyped. Although the connecting relatives have not been genotyped, since this is a rare variant, it is almost certain that they are "obligate" carriers of the variant. Quantitative cosegregation analysis of this family resulted in a likelihood ratio of 28:1 which

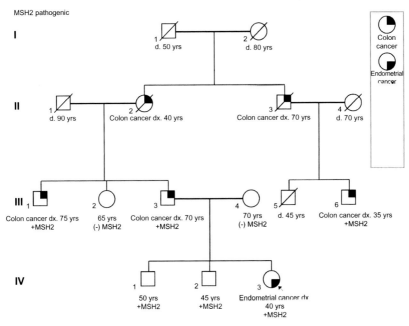

FIG. 7 Example pedigree showing co-segregation of a pathogenic variant in *MSH2*.

is strong evidence supporting the hypothetical MSH2 variant in this family causing Lynch syndrome.

2. *BRCA2* variant with cosegregation evidence against pathogenicity (Fig. 8). When a 55-year-old woman with breast cancer has a *BRCA2* variant and a family history of breast and ovarian cancer, the variant is often assumed to be the explanation for the family's multiple incidences of cancer. In this family, analysis showed that the maternal aunt with ovarian cancer did not have the variant. A cousin who had breast cancer did not have the variant. Two unaffected relatives who had the variant revealed multiple deceased relatives who were obligate carriers (individuals who must also have the variant if we assume it is inherited from a common ancestor) and did not have cancer. The cosegregation likelihood ratio for this variant in this family was 0.22:1 which is moderate evidence suggesting this variant is benign.

3. A small family with an *ATM* variant (Fig. 9). Many modern families are small. This family shows a 50-year-old woman with breast cancer who carries an *ATM* variant. Individuals with two pathogenic variants in *ATM* have the recessively inherited ataxia-telangectasia syndrome. Women with one pathogenic *ATM* variant have 2 to 3-fold risk of breast cancer. This small pedigree shows that this woman inherited her *ATM* variant from her mother, who is 75 and has not had breast cancer. Knowing this lowers the likelihood that the specific variant in this family causes a moderately increased risk for breast cancer. The cosegregation likelihood

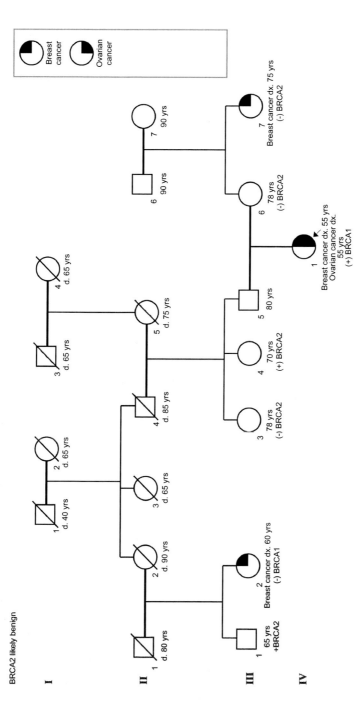

FIG. 8 Example pedigree showing a benign variant in *BRCA2* that does not cosegregate with breast cancer.

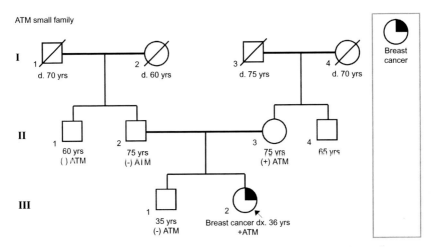

ATM small family

FIG. 9 Example pedigree showing a variant in the *ATM* gene with minimal evidence from cosegregation analysis.

ratio from this family is 0.82:1. Information from other sources or additional families with the same ATM variant would need to be combined with this information to classify the pathogenicity of this *ATM* variant.

4. Connecting distant relatives with the same *PMS2* variant (Fig. 10A–C). There were two patients that each independently found that they had the same *PMS2* variant. One man had colorectal cancer at 50, and a woman had endometrial cancer at 35 (Fig. 10A). The man found out that his sister, who also had colorectal cancer at 45 had the same *PMS2* variant. His other sister that decided to be tested did not have the *PMS2* variant. The cosegregation likelihood ratio for this small family was 1.8:1 which is modest evidence to support pathogenicity, but not sufficient to reclassify the variant.

In the other family the woman with endometrial cancer knew that both of her paternal grandparents had developed colorectal cancer, and she was able to determine that she inherited the *PMS2* variant from her father even though both her grandparents were deceased (Fig. 10B). Her younger sisters both had the same *PMS2* variant. The cosegregation likelihood ratio for her family was 1.57:1.

Neither of these families had a dramatic cancer history, but *PMS2* variants generally have a lower penetrance than other Lynch syndrome genes, so it is not unusual to see PMS2 variant carriers who never have cancer. Each of the two index patients happened to do DNA testing to find out about their family history, and found out that they were cousins. They contacted each other to see if they could help each other out with their genealogy. After a little work they found that they were 2nd cousins once removed. During their conversations about family history they discovered that they were both cancer survivors, and more surprisingly that they each had the same *PMS2* VUS. During their family history research they also found out that one of their great aunts died of endometrial cancer at the age of 50. Each of the two small pedigrees had a

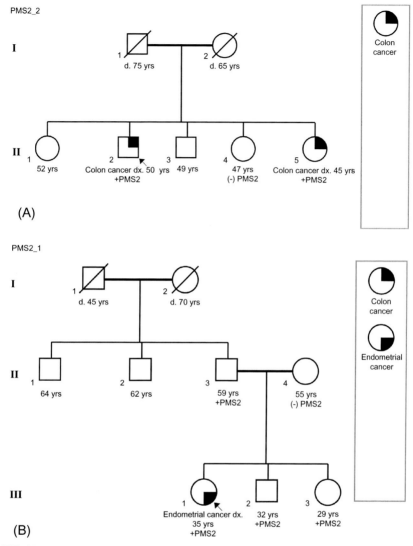

FIG. 10 (A) Example of a small pedigree showing modest evidence of cosegregation of a variant in the *PMS2* gene. (B) Example of a separate small pedigree showing the same variant as 10a in the *PMS2* gene with modest evidence of cosegregation.

(Continued)

PMS2 combined

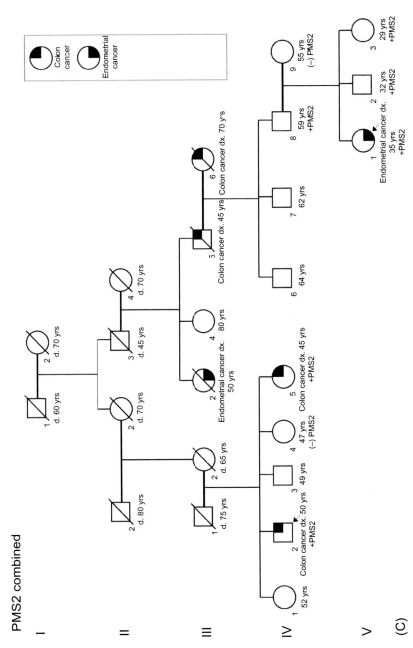

FIG. 10, CONT'D (C) Combining both pedigrees from (A) and (B) results in a larger pedigree with strong evidence supporting classification of the PMS2 variant as pathogenic.

modest cosegregation likelihood ratio. However, the combined pedigree yielded a likelihood ratio of 53:1, which is strong evidence indicating that the variant is very likely pathogenic (Fig. 10C). This information will be useful to these individual's living siblings and children who have the same *PMS2* variant as they can begin aggressive colonoscopy screening and may consider prophylactic hysterectomy after they have finished having children.

A major part of the cosegregation calculation is determining the probability that the individuals with the variant have inherited it by chance. Recall that the degree of relationship (e.g., first degree, second degree, etc.) between family members was delineated by noting the number of intervening meiotic events (and resulting births) between two relatives. For distant relations separated by many meiotic events there is a lower probability that they just happen to have the same variant by chance and therefore, give more information through cosegregation analysis.

This example is a unique situation that may become more common in the 21st century. Several research studies have identified ancestral connections between participants with the same rare variant. The large pedigrees created by connecting individuals with the same variant are ideal for variant classification. Most previous studies of genetic cancer-risk communication among families have focused on the relationships and choices of individuals that know their genetic status with those that do not. Connecting putatively related individuals with the same rare VUS may create a new dynamic with new categories of participants.

In addition to cosegregation analysis, families can also provide many other types of data, which are more intuitive and could be gathered by an interested primary care physician. Data on parent of origin can reveal if a variant is pathogenic or benign (haplotype phasing configuration and de novo status) and imprinting as well as providing a sample of individuals highly enriched for the variant to ascertain basic genotype-phenotype correlation. Some examples of each of these situations are described below.

Examples of variant data that can be obtained through family studies beyond cosegregation

1. *De novo information*: A unique *SMAD4* variant was seen in a young father who had had blood in his stool, and who was subsequently found to have several juvenile polyps on colonoscopy. *SMAD4* is a gene associated with familial juvenile polypsis, however, the patient did not have any family history of colorectal cancer, despite having many aunts and uncles on both maternal and paternal sides. Both his mother and father had negative colonoscopies. It was considered likely that the variant was benign, and that there was another cause for the juvenile polyps because of the lack of family history. However, both parents were genotyped and found to be negative for the *SMAD4* variant. After paternity was confirmed, the *SMAD4* variant was classified as a likely pathogenic variant that was de novo in the patient.

Approximately 1% of clinically identified variants that are not in public databases are de novo. Sometimes de novo mutations are mosaic, or present in only a portion of the patient's cells. In this family the finding of a de novo variant is entirely consistent with the given family history. Siblings may have some risk of having the variant, if it initially occurred in the gonads of one of the patient's parents. Children of the patient may benefit from genetic testing when they are old enough that they may become symptomatic. Sperm testing on the patient might help clarify the risk that additional offspring will inherit his *SMAD4* variant.

2. *Haplotype phasing information*: A new *MUTYH* VUS and a pathogenic *MUTYH* variant were both seen in a woman with several polyps on a routine colonoscopy at age 50. One of the patient's 6 siblings also had polyps on his colonoscopy and another had colon cancer in her 50s. *MUTYH* causes an autosomal recessive polyposis syndrome. It is not clear from the patient testing if her two *MUTYH* variants are on the same copy of the gene (in *cis*) or on different copies (in *trans*). Testing the parents would resolve this. If one parent was found to have both the pathogenic variant and the VUS, it would be evidence that the VUS is benign, as that would prove the VUS to be on the same strand of DNA as the pathogenic variant (in *cis*). On the other hand, if the VUS and the pathogenic variant were seen in different parents, it would be evidence that the *MUTYH* variant is pathogenic. Testing siblings may also help phase variants, if only one of the variants is seen in a sibling it would indicates that each *MUTYH* variant is on a different allele (in *trans*).

3. *Imprinting information*: Imprinting is an epigenetic mechanism (see Chapter 5 and Chapter 9 for more on epigenetics) where either the maternal or paternal DNA copy is methylated, so the only functional copy of a gene is inherited from one parent. In medical school this is usually discussed using Prader-Willi and Angelman syndrome examples which are associated with maternally and paternally imprinted genes on chromosome 15. If a variant is in an imprinted gene, knowing if there is maternal or paternal inheritance can help classify the VUS. For example: a VUS in *CDKN1C* was identified in a man on a very broad genetic panel performed for hereditary cancer risk. *CDKN1C* is a paternally imprinted gene, so maternally inherited pathogenic variants have been reported to cause Beckwith-Wiedemann syndrome or IMAGe syndrome. The proband's father tested negative for the *CDKN1C* VUS, indicating probable maternal inheritance of the variant. Since neither the patient nor any of his siblings had any features of Beckwith-Wiedemann syndrome or IMAGe syndrome, the variant can be classified as likely benign. If the variant had been paternally inherited the family would not be as informative for the variant.

4. *Phenotype-genotype correlation*: A missense variant was identified in the *APC* gene in a 37-year-old woman who had a hereditary cancer panel test after being diagnosed with breast cancer. It was initially classified as a VUS. The *APC* variant was in the last half of the gene. Pathogenic nonsense and

frameshift mutations in the last half of the APC gene cause an attenuated form of familial adenomatous polyposis (FAP), which can cause 10–100 polyps and very high lifetime risk of colon cancer.[59] If this variant were pathogenic, it would be predicted to cause this attenuated FAP. Insurance would not pay for her to have a colonoscopy, as there was no family history of early onset colon cancer; however, her family was small. The patient's parents were both genotyped. Her 80-year-old father was found to have the same *APC* variant and had recently received his second clean colonoscopy. The *APC* variant was classified as likely benign. In silico predictions suggested no change in protein, and it would be extremely unlikely for an 80-year-old with a polyposis-causing variant to have a clean colonoscopy. In situations like this, it is impossible to rule out the possibility that the *APC* variant might confer a very low risk of polyps as polyps are only ever seen in a portion of the general population. However, we can rule out a high risk of polyps and colorectal cancer. Less than a 2-fold risk of colorectal cancer and polyps may not justify screening beyond the screening colonoscopies recommended for the general population.

Conclusion: What to do when with a patient has a genetic variant that you have never seen before

In the age of genetic testing it is almost certain that you will increasingly begin to encounter patients who want to talk to a physician about the family-specific genetic variant they have identified through any number of genetic testing options. Generally, if the variant is listed as pathogenic or likely pathogenic the medical management implications will be described by medical specialty guidelines and outlined in the test report. If the variant is listed as a variant of uncertain significance, the consensus is that medical management should be based on personal and family history. **Treatment should not be based on VUS findings.**

You may be able to help your patient classify her VUS. The first, and probably most important thing you can do is take a complete family history and draw a pedigree. Make sure your patient has a copy of this. Contact the laboratory that performed the genetic test to determine if any family VUS classification studies are available for your patient. Sometimes these family studies are offered for free and other times there may be a fee for testing additional relatives.

There are many national programs and international databases interested in variant classification. As mentioned above, ClinVar aggregates data from clinical and research laboratories.[14] ClinGen curates this data.[45] PROPMT is a registry for patients who wish to share their genetic findings with researchers.[60] There are a growing number of organizations focused on understanding variants in specific genes and diseases. InSiGHT focuses on Lynch syndrome genes, *MLH1*, *MSH2*, *MSH6*, and *PMS2* as well as other colorectal cancer risk genes.[61] ENIGMA focuses on hereditary breast cancer genes *BRCA1* and *BRCA2*.[62] The landscape of patient advocacy programs is too long and dynamic to list in a book, a disease specific internet search for advocacy programs is likely to yield several relevant results.

Some patients understand and are completely fine with genetic uncertainty. In other patients uncertain findings cause anxiety and lead to inappropriate care. All patients, and specifically patients that are anxious about uncertain genetic findings, may benefit from based family efforts that help them identify and communicate with relatives that have the same genetic variants.

Summary and conclusions

Because dozens of new genetic variants enter the human population with each new birth, every genetic variant that is compatible with life is present in the human population today. Most of these could be termed family-specific variants because they cluster in extended families rather than being distributed evenly throughout the population. Family-specific variants can are often described using a 5-level scale that classifies variants as benign, likely benign, variant of uncertain significance, likely pathogenic or pathogenic. Even though most family-specific variants are benign, a small fraction of all family-specific variants will be classified as pathogenic. These pathogenic family-specific variants cause the majority of dominant adult onset disease. Many family-specific variants are initially classified as variants of uncertain significance until enough information can be gathered to determine if the variants are associated with disease or not. Family studies that use cosegregation analysis to determine if the variant travels with disease in the family more often than expected by chance are one of the most efficient ways reclassify variants of uncertain significance.

- If your patients have genetic sequencing test done, you are likely to see reports that list family-specific variants that you have never seen before.
- If a variant is currently classified as a variant of uncertain significance it is mostly likely to be reclassified as benign. Current guidelines suggest that the presence of a VUS should not alter clinical care.
- Computational, functional, and clinical information from patients and their families can be used to reclassify VUS.
- Family studies are one of the best ways to gather information to classify variants of uncertain significance.

Brian H. Shirts is the lead author for this chapter. First person statements are from his point of view.

References

1. Jarvik GP, Evans JP. Mastering genomic terminology. *Genet Med.* 2017;19(5):491–492. https://doi.org/10.1038/gim.2016.139.
2. Narasimhan VM, Rahbari R, Scally A, et al. Estimating the human mutation rate from autozygous segments reveals population differences in human mutational processes. *Nat Commun.* 2017;8(1):303. https://doi.org/10.1038/s41467-017-00323-y.

3. Palamara PF, Francioli LC, Wilton PR, et al. Leveraging distant relatedness to quantify human mutation and gene-conversion rates. *Am J Hum Genet*. 2015;97(6):775–789. https://doi.org/10.1016/j.ajhg.2015.10.006.

4. Delatycki MB, Wolthuizen M, Aitken MA, Hickerton C, Metcalfe SA, Allen KJ. To tell or not to tell—what to do about p.c282Y heterozygotes identified by HFE screening. *Clin Genet*. 2013;84(3):286–289. https://doi.org/10.1111/cge.12053.

5. Laberge A-M. Recommending inclusion of HFE C282Y homozygotes in the ACMG actionable gene list: cop-out or stealth move toward population screening? *Genet Med*. 2018;20(4):400–402. https://doi.org/10.1038/gim.2017.161.

6. Yapp TR, Eijkelkamp EJ, Powell LW. Population screening for HFE-associated haemochromatosis: should we have to pay for our genes? *Intern Med J*. 2001;31(1):48–52.

7. Karami F, Mehdipour P. A comprehensive focus on global spectrum of BRCA1 and BRCA2 mutations in breast cancer. *Biomed Res Int*. 2013;2013:928562. https://doi.org/10.1155/2013/928562.

8. Levy-Lahad E, Catane R, Eisenberg S, et al. Founder BRCA1 and BRCA2 mutations in Ashkenazi Jews in Israel: frequency and differential penetrance in ovarian cancer and in breast-ovarian cancer families. *Am J Hum Genet*. 1997;60(5):1059–1067.

9. Oddoux C, Struewing JP, Clayton CM, et al. The carrier frequency of the BRCA2 6174delT mutation among Ashkenazi Jewish individuals is approximately 1%. *Nat Genet*. 1996;14(2):188–190. https://doi.org/10.1038/ng1096-188.

10. Haraldsdottir S, Rafnar T, Frankel WL, et al. Comprehensive population-wide analysis of Lynch syndrome in Iceland reveals founder mutations in MSH6 and PMS2. *Nat Commun*. 2017;8:14755. https://doi.org/10.1038/ncomms14755.

11. Clendenning M, Senter L, Hampel H, et al. A frame-shift mutation of PMS2 is a widespread cause of Lynch syndrome. *J Med Genet*. 2008;45(6):340–345. https://doi.org/10.1136/jmg.2007.056150.

12. Ranola JMO, Pearlman R, Hampel H, Shirts BH. Modified capture-recapture estimates of the number of families with Lynch syndrome in Central Ohio. *Familial Cancer*. 2019;18(1):67–73. https://doi.org/10.1007/s10689-018-0096-0.

13. Clendenning M, Baze ME, Sun S, et al. Origins and prevalence of the American Founder Mutation of MSH2. *Cancer Res*. 2008;68(7):2145–2153. https://doi.org/10.1158/0008-5472.CAN-07-6599.

14. Landrum MJ, Lee JM, Riley GR, et al. ClinVar: public archive of relationships among sequence variation and human phenotype. *Nucleic Acids Res*. 2014;42(Database issue):D980–D985. https://doi.org/10.1093/nar/gkt1113.

15. Lek M, Karczewski KJ, Minikel EV, et al. Analysis of protein-coding genetic variation in 60,706 humans. *Nature*. 2016;536(7616):285–291. https://doi.org/10.1038/nature19057.

16. Macklin S, Durand N, Atwal P, Hines S. Observed frequency and challenges of variant reclassification in a hereditary cancer clinic. *Genet Med*. 2018;20(3):346–350. https://doi.org/10.1038/gim.2017.207.

17. Mersch J, Brown N, Pirzadeh-Miller S, et al. Prevalence of variant reclassification following hereditary cancer genetic testing. *JAMA*. 2018;320(12):1266–1274. https://doi.org/10.1001/jama.2018.13152.

18. Amendola LM, Jarvik GP, Leo MC, et al. Performance of ACMG-AMP variant-interpretation guidelines among nine laboratories in the clinical sequencing exploratory research consortium. *Am J Hum Genet*. 2016;https://doi.org/10.1016/j.ajhg.2016.03.024.

19. Harrison SM, Dolinksy JS, Chen W, et al. Scaling resolution of variant classification differences in ClinVar between 41 clinical laboratories through an outlier approach. *Hum Mutat*. 2018;39(11):1641–1649. https://doi.org/10.1002/humu.23643.

20. Eggington JM, Bowles KR, Moyes K, et al. A comprehensive laboratory-based program for classification of variants of uncertain significance in hereditary cancer genes. *Clin Genet.* 2013;8(10):12315.

21. Makhnoon S, Garrett LT, Burke W, Bowen DJ, Shirts BH. Experiences of patients seeking to participate in variant of uncertain significance reclassification research. *J Community Genet.* 2018;https://doi.org/10.1007/s12687-018-0375-3.

22. Plon SE, Eccles DM, Easton D, et al. Sequence variant classification and reporting: recommendations for improving the interpretation of cancer susceptibility genetic test results. *Hum Mutat.* 2008;29(11):1282–1291. https://doi.org/10.1002/humu.20880.

23. Shirts BH, Pritchard CC, Walsh T. Family-specific variants and the limits of human genetics. *Trends Mol Med.* 2016;.

24. Vos J, Otten W, van Asperen C, Jansen A, Menko F, Tibben A. The counsellees' view of an unclassified variant in BRCA1/2: recall, interpretation, and impact on life. *Psychooncology.* 2008;17(8):822–830.

25. Richards S, Aziz N, Bale S, et al. Standards and guidelines for the interpretation of sequence variants: a joint consensus recommendation of the American College of Medical Genetics and Genomics and the Association for Molecular Pathology. *Genet Med.* 2015;5(10):30.

26. Makhnoon S, Shirts BH, Bowen DJ, Fullerton SM. Hereditary cancer gene panel test reports: wide heterogeneity suggests need for standardization. *Genet Med.* 2018;20(11):1438–1445. https://doi.org/10.1038/gim.2018.23.

27. O'Neill SC, DeMarco T, Peshkin BN, et al. Tolerance for uncertainty and perceived risk among women receiving uninformative BRCA1/2 test results. *Am J Med Genet C Semin Med Genet.* 2006;142C(4):251–259.

28. O'Neill SC, Rini C, Goldsmith RE, Valdimarsdottir H, Cohen LH, Schwartz MD. Distress among women receiving uninformative BRCA1/2 results: 12-month outcomes. *Psychooncology.* 2009;18(10):1088–1096. https://doi.org/10.1002/pon.1467.

29. van Dijk S, Timmermans DR, Meijers-Heijboer H, Tibben A, van Asperen CJ, Otten W. Clinical characteristics affect the impact of an uninformative DNA test result: the course of worry and distress experienced by women who apply for genetic testing for breast cancer. *J Clin Oncol.* 2006;24(22):3672–3677.

30. Culver J, Brinkerhoff C, Clague J, et al. Variants of uncertain significance in BRCA testing: evaluation of surgical decisions, risk perception, and cancer distress. *Clin Genet.* 2013;17(10):12097.

31. Garcia C, Lyon L, Littell RD, Powell CB. Comparison of risk management strategies between women testing positive for a BRCA variant of unknown significance and women with known BRCA deleterious mutations. *Genet Med.* 2014;16(12):896–902. https://doi.org/10.1038/gim.2014.48.

32. Murray ML, Cerrato F, Bennett RL, Jarvik GP. Follow-up of carriers of BRCA1 and BRCA2 variants of unknown significance: variant reclassification and surgical decisions. *Genet Med.* 2011;13(12):998–1005. https://doi.org/10.1097/GIM.0b013e318226fc15.

33. Boesen M, Writer G. *I Got A Double Mastectomy After A Genetic Test. Then I Learned The Results Were Wrong.* (500, 00:01). Available from: https://www.huffpost.com/entry/brca-genetic-testing-mastectomy_n_5c6c39fbe4b012225acd80d3; 2019. (Accessed March 6, 2019).

34. Oregon Lawsuit Highlights Importance of Genetic Counseling During Period of Increasing Test Access. (n.d.). Available from: https://www.genomeweb.com/cancer/oregon-lawsuit-highlights-importance-genetic-counseling-during-period-increasing-test-access Accessed 6 March 2019

35. Macklin S, Durand N, Atwal P, Hines S. Observed frequency and challenges of variant reclassification in a hereditary cancer clinic. *Genet Med.* 2018;20(3):346–350. https://doi.org/10.1038/gim.2017.207.

36. Tsai GJ, Rañola JMO, Smith C, et al. Outcomes of 92 patient-driven family studies for reclassification of variants of uncertain significance. *Genet Med.* 2018;21:1435–1442. https://doi.org/10.1038/s41436-018-0335-7.

37. Garrett LT, Hickman N, Jacobson A, et al. Family studies for classification of variants of uncertain classification: current laboratory clinical practice and a new web-based educational tool. *J Genet Couns.* 2016;25(6):1146–1156. https://doi.org/10.1007/s10897-016-9993-2.

38. Tonelli MR, Shirts BH. Knowledge for precision medicine: mechanistic reasoning and methodological pluralism. *JAMA.* 2017;318(17):1649–1650. https://doi.org/10.1001/jama.2017.11914.

39. Ghosh R, Harrison SM, Rehm HL, Plon SE, Biesecker LG, ClinGen Sequence Variant Interpretation Working Group. Updated recommendation for the benign stand-alone ACMG/AMP criterion. *Hum Mutat.* 2018;39(11):1525–1530. https://doi.org/10.1002/humu.23642.

40. Li Q, Wang K. InterVar: clinical interpretation of genetic variants by the 2015 ACMG-AMP guidelines. *Am J Hum Genet.* 2017;100(2):267–280. https://doi.org/10.1016/j.ajhg.2017.01.004.

41. Niehaus A, Azzariti DR, Harrison SM, et al. A survey assessing adoption of the ACMG-AMP guidelines for interpreting sequence variants and identification of areas for continued improvement. *Genet Med.* 2019;21(8):1699–1701. https://doi.org/10.1038/s41436-018-0432-7.

42. Nykamp K, Anderson M, Powers M, et al. Sherloc: a comprehensive refinement of the ACMG-AMP variant classification criteria. *Genet Med.* 2017;19(10):1105–1117. https://doi.org/10.1038/gim.2017.37.

43. Tavtigian SV, Greenblatt MS, Harrison SM, et al. Modeling the ACMG/AMP variant classification guidelines as a Bayesian classification framework. *Genet Med.* 2018. https://doi.org/10.1038/gim.2017.210.

44. Rivera-Muñoz EA, Milko LV, Harrison SM, et al. ClinGen variant curation expert panel experiences and standardized processes for disease and gene-level specification of the ACMG/AMP guidelines for sequence variant interpretation. *Hum Mutat.* 2018;39(11):1614–1622. https://doi.org/10.1002/humu.23645.

45. ClinGen. (n.d.). Available from: http://www.iccg.org/about-the-iccg/clingen/

46. Dong C, Wei P, Jian X, et al. Comparison and integration of deleteriousness prediction methods for nonsynonymous SNVs in whole exome sequencing studies. *Hum Mol Genet.* 2015;24(8):2125–2137. https://doi.org/10.1093/hmg/ddu733.

47. Li J, Zhao T, Zhang Y, et al. Performance evaluation of pathogenicity-computation methods for missense variants. *Nucleic Acids Res.* 2018;46(15):7793–7804. https://doi.org/10.1093/nar/gky678.

48. Mather CA, Mooney SD, Salipante SJ, et al. CADD score has limited clinical validity for the identification of pathogenic variants in noncoding regions in a hereditary cancer panel. *Genet Med.* 2016;18:1269–1275. https://doi.org/10.1038/gim.2016.44.

49. Starita LM, Ahituv N, Dunham MJ, et al. Variant interpretation: functional assays to the rescue. *Am J Hum Genet.* 2017;101(3):315–325. https://doi.org/10.1016/j.ajhg.2017.07.014.

50. Clark MB, Mercer TR, Bussotti G, et al. Quantitative gene profiling of long noncoding RNAs with targeted RNA sequencing. *Nat Methods.* 2015;12(4):339–342. https://doi.org/10.1038/nmeth.3321.

51. Mercer TR, Clark MB, Crawford J, et al. Targeted sequencing for gene discovery and quantification using RNA CaptureSeq. *Nat Protoc.* 2014;9(5):989–1009. https://doi.org/10.1038/nprot.2014.058.

52. Teng J, Risch N. The relative power of family-based and case-control designs for linkage disequilibrium studies of complex human diseases. II. Individual genotyping. *Genome Res.* 1999;9(3):234–241.

53. Thornton T, McPeek MS. Case-control association testing with related individuals: a more powerful quasi-likelihood score test. *Am J Hum Genet*. 2007;81(2):321–337. https://doi.org/10.1086/519497.

54. Shirts BH, Jacobson A, Jarvik GP, Browning BL. Large numbers of individuals are required to classify and define risk for rare variants in known cancer risk genes. *Genet Med*. 2013;19(10):187. https://doi.org/10.1038/gim.2013.187.

55. Thompson D, Easton DF, Goldgar DE. A full-likelihood method for the evaluation of causality of sequence variants from family data. *Am J Hum Genet*. 2003;73(3):652–655. https://doi.org/10.1086/378100.

56. Jarvik GP, Browning BL. Consideration of cosegregation in the pathogenicity classification of genomic variants. *Am J Hum Genet*. 2016;98(6):1077–1081. https://doi.org/10.1016/j.ajhg.2016.04.003.

57. Rañola JMO, Liu Q, Rosenthal EA, Shirts BH. A comparison of cosegregation analysis methods for the clinical setting. *Familial Cancer*. 2018;17(2):295–302. https://doi.org/10.1007/s10689-017-0017-7.

58. Gong G, Hannon N, Whittemore AS. Estimating gene penetrance from family data. *Genet Epidemiol*. 2010;34(4):373–381. https://doi.org/10.1002/gepi.20493.

59. Ibrahim A, Barnes DR, Dunlop J, Barrowdale D, Antoniou AC, Berg JN. Attenuated familial adenomatous polyposis manifests as autosomal dominant late-onset colorectal cancer. *Eur J Hum Genet*. 2014;22(11):1330–1333. https://doi.org/10.1038/ejhg.2014.20.

60. Hereditary Cancer Risks & Multiplex Gene Panels. (n.d.). Available from: http://promptstudy.info/ Accessed 6 March 2019

61. Plazzer JP, Sijmons RH, Woods MO, et al. The InSiGHT database: utilizing 100 years of insights into Lynch syndrome. *Familial Cancer*. 2013;12(2):175–180. https://doi.org/10.1007/s10689-013-9616-0.

62. Spurdle AB, Healey S, Devereau A, et al. ENIGMA – evidence-based network for the interpretation of germline mutant alleles: an international initiative to evaluate risk and clinical significance associated with sequence variation in BRCA1 and BRCA2 genes. *Hum Mutat*. 2012;33(1):2–7. https://doi.org/10.1002/humu.21628.

Chapter 5

Genes and cancer: Implications for FHH analysis

Vincent C. Henrich, Lori A. Orlando, and Brian H. Shirts

- The risk for hereditary cancer onset depends on how cell growth regulation is impaired: The *BRCA1/2* example.
- Hereditary non-polyposis colorectal cancer (HNPCC, aka Lynch syndrome) also involves variants in genes necessary for DNA repair.
- A pathogenic variant increases a carrier's risk for HBOC and Lynch syndrome, but onset requires a second environmentally-induced event.
- Different cancers in a family can result from the same genetic variant.
- Early age of onset, severity, and recurrence are evidence of a hereditary cancer syndrome.
- Cancer occurrence in second degree relatives may be necessary to diagnose hereditary syndromes like HBOC or Lynch syndrome.
- Diagnosing a hereditary cancer syndrome sometimes requires simultaneous consideration of several factors.
- Family health history can provide insight about complex gene-environment interactions: pancreatic cancer.
- Genetic biomarkers could have utility for refining personal and family-related cancer risk assessment.
- Cancer occurrence sometimes involves non-genetic cellular mechanisms.
- Some hereditary cancers result from genetic variants that do not require an environmental event.
- Clinical measures of cancer risk and environmental risk factors can complicate cancer risk assessment.
- Summary and conclusions.

When contemplating the vulnerability that underlies a positive family health history for an adult onset disease, it is clear that only a fraction of the risk is attributable to genetic mutations. Those that are well known are mostly older variants that have been around thousands of years and are thus easily detectable; but there are some that have been around less long (centuries) that are more easily seen in smaller, more restricted, populations. To these are added new variants every day, since every human birth generates dozens of new variants.

Managing Health in the Genomic Era. https://doi.org/10.1016/B978-0-12-816015-2.00005-5

Most remain undetected and uncharacterized, making it difficult for genetic tests to elucidate a clear-cut genetic diagnosis.

The purpose of this chapter is to explore how family health history can be used to refine hereditary cancer risk, even in the absence of genetic testing or a meaningful genetic test result. Compiling and analyzing family health histories can be systematized to some degree by understanding the processes that are responsible for converting an inherent cancer risk into cancer onset. These processes follow from principles of cell biology and are manifested in ways that can be recognized through the family health history. There is a tendency to think of disease risk as comprised of completely separate genetic and environmental components. In fact, the interaction between the two is the major determinant for cancer incidence.

A major part of this discussion will focus on *BRCA1* and *BRCA2*, which are the most well-described of the genes associated with hereditary breast/ovarian cancer (HBOC). These genes also serve as a basis for understanding the relationship between a family's risk and cancer incidence. *BRCA1* and *BRCA2* also illustrate several principles that are important for assessing a family's cancer risk. Well over 1000 variants in these genes provide a strong indication about a patient's risk for developing cancer. Though, several confer an exceptionally high risk of cancer, with a penetrance as high as 75%.[1] But even these, the strongest of the pathogenic variants, do *not* lead to cancer in 25% of variant carriers. Thus, pathogenic *BRCA1* and *BRCA2* variants are predisposing, but not directly causative, and this has important implications for managing the patient. Why are some this high and others not? Why is it not 100%? We will explore these questions and more in this chapter. Many of the mechanisms associated with hereditary cancers involve principles that apply broadly, to familial patterns of disease. Given the seriousness and prevalence of cancer in the general population, the hereditary cancers serve as a useful starting point for exploring these principles.

The major question that will be addressed here is: Why do 25% of carriers for highly penetrant and known pathogenic *BRCA1* and *BRCA2* variants never develop cancer? Does this tell us something significant about assessing family health history for cancer prevention? How generally applicable are the insights about "incomplete penetrance" for managing patient health? The answers to these questions depend upon careful scrutiny of what a "cancer gene" normally does in cells, how a variant might disrupt that role, and the subsequent pathway of molecular and cellular events that eventually could lead to adult disease onset in *some* variant carriers.

The ensuing discussion will focus primarily on the genetic and cellular events that are relevant for understanding common hereditary cancer syndromes for breast/ovarian cancer and hereditary colorectal cancer. These syndromes occur frequently enough in any population that most providers are likely to encounter them repeatedly during their career. The challenge for the provider lies in identifying key features of a patient's family health history *early enough* to make a definitive risk assessment and offer family members specific diagnosis and treatment options for prevention, or at least, early detection of cancer. Moreover, the principles of genetics and cell biology underlying these hereditary cancer syndromes can be extended and applied to other cancer-predisposing genes. As will be seen later, genetic test results sometimes increase the certainty of risk assessments for carriers and noncarriers of relevant genetic variants for cancer, but since over 80% of clinical genetic testing for hereditary cancer yields negative test results, family health history provides the necessary and critical information for taking meaningful medical action.[2-8] Practically speaking, family health history is important—*especially* when a genetic test result cannot be obtained, or fails to provide clarity about a patient's risk. Unfortunately, the full value of family health history is often unrecognized, without some appreciation for what it reveals about a family's health risks and the considerations that are essential to remember when considering the possibility of an elevated cancer risk in the family.

Family health history raises awareness about factors that *precede* actual disease onset and allows the possibility for preventive interventions to bend the probability back towards maintaining health over developing cancer. The course of a cancer progression pathway conceptually follows a series of "forks" in the road to disease onset. Assigning risk to a patient is based on estimating the relative probabilities of events that forestall or predispose further progression towards a disease state. This is particularly relevant for families, since family members often share not only an inherited risk for a disease, but also share environmental exposures that can promote or prevent disease onset. Herein lies the opportunity to maintain good health, by identifying interventions that successfully mitigate their shared risks, and potentially their family-specific vulnerability.

The diagnosis of hereditary cancer syndromes, like Lynch and HBOC, illustrate how the pathway from genetic variation to a change in health status is not a one-step, cause-effect process. Typically, family health history provides the "first clues" for identifying hereditary cancer syndromes. The discovery of which is critically important not only for the provider and the at-risk patient, but also for their family members.

The risk for hereditary cancer onset depends on how cell growth regulation is impaired: The *BRCA1/BRCA2* example

The approach to diagnosis and management of an at-risk patient for familial diseases and conditions is enhanced by understanding relevant cellular and genetic principles. A pathogenic genetic variant typically alters or impairs a normal cellular process, but the effect may be more like an automobile engine that loses its optimum performance capabilities, sputtering along without observable difference to the driver, until encountering an unexpected demand, such as a steep hill. Similarly, an alteration of cellular function sets off a progression of changes that could eventually result in disease onset during the individual's lifetime.

The relevance of the relationship between genetics, cell biology and disease progression, is illustrated by the interplay of the *BRCA1* and *BRCA2* genes with cellular activity, and their connection to HBOC. The central importance of the *BRCA1* and *BRCA2* gene variants is their *normal* role, to prevent cancer. Literally, the normal work of these two proteins is essential for regulating normal cell growth cycles in breast and other tissues. The *BRCA1* and *BRCA2* genes are the blueprints for two proteins that repair DNA damaged by environmental hazards, notably ionizing radiation. Since exposure to environmental radiation is inevitable, without a mechanism for repairing damaged DNA, cancer would be inevitable, as well. In addition, other environmental exposures, can sometimes further promote the likelihood of developing cancer in those who are already genetically predisposed. It is important to note that "environmental" triggers also exert their effects by affecting proteins whose general role is to restore normal cell function (i.e., **homeostasis**). In other words, both genetic mutations and environmental hazards cause cancer by damaging cell's DNA and inhibiting normal cellular functions.

So, what does the BRCA1 protein normally do? How do pathogenic variants affect its normal role? Does understanding these cellular functions provide insights about cancer risk and prevention? BRCA1 is a protein involved in correcting errors that arise periodically in any actively dividing cell. Faithful replication of a new DNA copy from a cell's existing one is essential. As a cell grows, it divides, and each daughter cell must receive an exact DNA copy (replication). Now, contemplate all the many daily environmental challenges encountered during a normal cell's life that can damage its DNA. Every one of us is continually bombarded with ultraviolet light, ionizing radiation, and highly reactive chemicals that we eat, drink, absorb, breathe, or sequester in our bodies. Even if DNA replication was an infallible process (and it's not), the possibility exists that a cell's DNA could sustain damage that results in a DNA change (pathogenic variant) that could, in turn, induce a loss of cellular regulation. But cells are not defenseless. Because the damage is potentially catastrophic to life processes, cells have evolved an elaborate repair mechanism for this kind of DNA damage.

Suppose DNA, in a cell that periodically divides as part of its normal life cycle, is hit by ionizing radiation resulting in breakage of double-stranded DNA somewhere along one of its 23 chromosome pairs. The event is rare in absolute terms, but because there are so many cells (about 37.2 trillion), some of them will repeatedly sustain this kind of damage over the course of a lifetime. When a chromosome's double-stranded DNA is broken by irradiation, the corresponding DNA sequence from the other chromosome (the **homologous DNA**—*remember everyone has 23* **pairs** *of chromosomes one from each parent*) serves as a template for resynthesizing the normal DNA sequence (Fig. 1). The process of **homology directed DNA repair** is far more complex than a few sentences

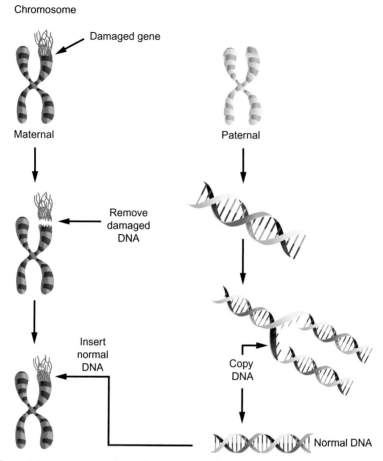

FIG. 1 Schematic diagram of homologous DNA repair. In this example of homologous repair the maternal chromosome has a damaged gene and the paternal chromosome serves as the donor of normal DNA. This, of course, can also happen in the reverse, with the mother's chromosome serving as the normal template for a damaged paternal chromosome.

allow and it involves the regulated interaction of numerous proteins dedicated to restoring the integrity of the DNA sequence—an indication of the crucial importance of the repair process.[9] Two essential molecular players for this repair process are the BRCA1 and BRCA2 proteins, which interact with each other and several other proteins to orchestrate a multistep and coordinated repair process. BRCA1 not only fulfills a central role in managing the homologous directed DNA repair process in response to damage, it also interacts with other proteins to stall the synthesis of new DNA until the cell's repair is complete. In addition, it influences the transcription of genes whose protein products normally play a role in the events that comprise cell division (chromosome condensation, mitotic division, and cytokinesis). One of the proteins that is specifically tied to the DNA repair process itself is BRCA2. The inevitable need for DNA repair mechanisms in growing and dividing cells is particularly essential during a woman's reproductive age, because estrogen and other growth-stimulating hormones, promote cell proliferation, which increase the opportunities for life threatening damage to occur.

The protein-protein interactions of the repair proteins are not simply the consequence of random (Brownian) motion occurring in a restricted space. Rather, the contacts are highly ordered and involve very specific structures within the BRCA1 protein as well as the proteins with which it interacts. These specific functions within the BRCA1 protein define its *domains*, that is, the structural regions whose shape and charge characteristics are determined by the protein's amino acid sequence. These features allow BRCA1 to carry out its preordained functions. What does knowing about protein domains (and their amino acid sequence) mean, if anything, for medical practice? The answer is: domains allow a protein to take specific shapes and thus, interact with other molecules in the cell to perform specific cellular functions. The pathogenicity of a *BRCA1* variant sometimes results from disruption of important domains by a change in its DNA sequence. This is not the only mechanism for protein dysfunction to arise, but it is a common one.

Despite its seeming simplicity, the true picture of BRCA1 and BRCA2 protein function is even more dynamic. As with all genes the process of going from DNA to proteins involves several steps: the transcription of DNA to messenger RNA (mRNA) and translation of mRNA to protein. During the transcription process portions of the mRNA are "spliced" out yielding a final mRNA sequence that includes some but not all of the underlying DNA sequences. Depending upon the cell's current needs, different portions of the mRNA can be spliced out yielding slightly different forms of the protein that work similarly, but not exactly the same as each other (isoforms). This process is called alternative splicing and can be very important in increasing the diversity of cellular functions (Fig. 2). Given the importance of BRCA1 and its multiple roles in cell regulation, it has multiple alternative splices that result in several different BRCA1 protein isoforms. Frequently, different cell types will contain different isoforms, depending upon the cell's functional role, but there can also be a "mix" of isoforms within the same cell. So isoforms allow a single gene to fulfill different functions in different cells. This process explains how a pathogenic variant in a gene important for DNA repair in all cells, might preferentially lead to cancer in a specific tissue type, such as breast.

All BRCA1 isoforms share a common amino acid sequence, and due to alternative splicing, unique amino acid sequences that allow it to join with other proteins in unique protein complexes. One complex is critical for maintaining cellular integrity via homologous DNA repair, while another regulates the cell cycle, and a third controls "downstream" gene activity in the growing and dividing cell (Fig. 3). Importantly, the BRCA1 protein also interacts with several other proteins to *restore* homeostasis in the aftermath of a DNA-damaging

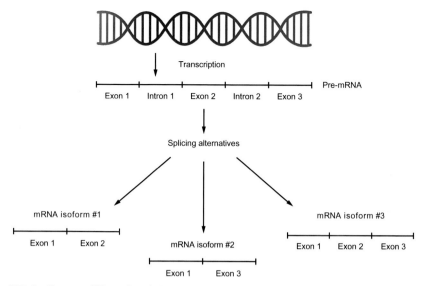

FIG. 2 Conceptual illustration of alternative splicing with three different isoforms for a hypothetical gene. A pre-mRNA is transcribed from the DNA. The final mRNA is spliced out and serves as a template for protein translation. In many cases, alternative splicing forms are possible as illustrated here.

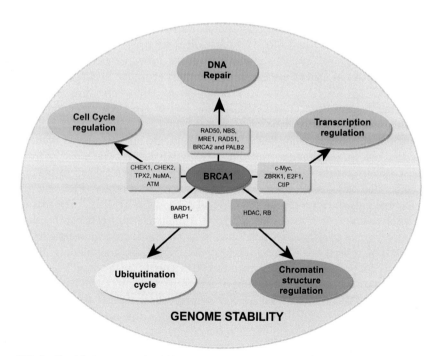

FIG. 3 Graphical representation of interactions between BRCA1 proteins and other cellular proteins involved in maintaining genomic stability as a result of damage to double-stranded DNA. BRCA1 isoforms (described in text) form protein complexes to carry out several molecular subfunctions necessary to restore genome stability in the aftermath of DNA damage.

event. The accompanying protein brigade includes ATM, CHEK2, PALB2, and BRCA2, all of which are linked to familial and hereditary cancers. In this book we will not list gene names, as it is not a genetics textbook, but they can be found online. As an illustration the protein, PALB2, stands for Partner and Locator of BRCA2, which defines its molecular role, and, also, implies how its dysfunction may affect cells. All of these genes have known variants that are moderately pathogenic for cancer, since all are necessary for homeostasis restoration to occur normally. They are also included on many cancer gene panels and are responsible for some of the HBOC cases.

Some of the most pathogenic genetic variants of *BRCA1* and *BRCA2* disrupt translation because the mRNA is damaged or unstable, creating a non-functional protein. In fact, the first three pathogenic founder variants discovered (remember Chapter 3's discussion on founder variants?) affect this pathway. The founders for these variants (all are short indels (insertions or deletions) though they are classified as SNPs; Table 1, Chapter 3) lived long ago in Ashkenazi Jewish populations. Since that time, the variants have spread as affected individuals have migrated. Importantly, even though these variants are highly pathogenic and occur at a younger age, they persist in the population because most cancers do not arise until after child bearing. It is worth keeping in mind that new *BRCA1/2* variants continuously arise with every birth, but new variants, even if they are highly pathogenic, will not be detected by genotyping (as discussed in Chapter 3). This is particularly relevant as direct to consumer companies, like 23&Me, offer *BRCA1* and *BRCA2* gene testing, but only perform genotyping on a few specific well-known variants (such as the 3 founder variants described above), which can lead to inaccurate expectations about the likelihood of having HBOC.

In the case of BRCA1 and BRCA2, the loss of the DNA repair function ensures accumulation of additional genetic errors and even cellular failure. Thus, it's not so surprising that *BRCA1* variants are also associated with an elevated risk for developing "triple negative" breast cancer, which results from the secondary genetic disruption of genes encoding three critical cell regulatory proteins: the estrogen receptor (ER), the progesterone receptor (PR), and the epidermal growth factor receptor (EGFR, also known as HER2).[10] The loss of all three receptors in a cancerous cell poses a challenge for treatment, because the repertoire of pharmaceutical treatments for breast cancer requires the intact presence of at least one of these receptors for therapeutic drug binding. If the protein is not present, it obviously can't serve as a target for the drug. Ironically, cancerous cells that have lost control of their growth due to secondary genetic damage, have a selective advantage over normal cells that retain the drug binding targets. The cancer cells are immune to the drug, while normal cells are exposed to its toxic effects, resulting in a steadily increasing proportion of cancer cells in the tissue.

The centrality of the BRCA1 and BRCA2 proteins for orchestrating the repair of DNA damage explains why many highly disruptive variants are highly

TABLE 1 Relative frequency of several cancer types in carriers of lynch syndrome genetic variants.

Cancer type	General population risk	MLH1 and MSH2		MSH6	PMS2
		Risk	Mean age of onset	Risk	Risk
Colorectal	5.5%	M: 27%–74% F: 22%–53%	27–46 yrs	M: 22% F: 10%	M: 20% F: 15%
Endometrial	2.7%	14%–54%	48–62 yrs	16%–26%	15%
Gastric	<1%	0.2%–13%	49–55 yrs	M: 6% F: 22%	6%
Ovarian	1.6%	4%–20%	43–45 yrs		
Small bowel	<1%	4%–12%	49 yrs		
Hepatobiliary tract	<1%	0.2%–4%	54–57		
Urinary tract	<1%	0.2%–25%	52–60 yrs		
Brain	<1%	1%–4%	~50 yrs		
Sebaceous neoplasms	<1%	1%–9%	Not reported	Unknown	Unknown
Pancreas	1.5%	0.4%–4%	63–65 yrs	Unknown	Unknown
Prostate	16.2%	9%–30%	59–60 yrs	Unknown	Unknown
Breast	12.4%	5%–18%	52 yrs	Unknown	Unknown

F, female; M, male.

Adapted from Adam MP, Ardinger HH, Pagon RA, et al., editors. GeneReviews® [Internet]. Seattle (WA): University of Washington, Seattle; 1993–2019. Available from: https://www.ncbi.nlm.nih.gov/books/NBK1116/. © University of Washington 1993–2017. GeneReviews® is the registered trademark of the University of Washington, Seattle. The pathogenic. By contrast, other variants exert little influence on the protein's functions and are, therefore, benign. This wide range of genetic effects illustrates a recurring principle discussed in Chapter 4: the medical impact of a gene's variants should be assessed by observing individuals who carry the variant. And where are such people most readily found? Among closely related family members.

From the medical management standpoint, the need for intensive risk management is evident when a pathogenic *BRCA1* variant is found, and prophylactic surgery to remove hormonally vulnerable tissue is recommended. However, as described in Chapter 4, low penetrance variants and VUSs complicate decisions around what the appropriate interventions are. Perhaps the greater challenge is determining whether a patient has a family health history that warrants closer scrutiny for the possibility of a hereditary cancer syndrome. How to implement complex family history assessments in clinical practice are discussed later in the book.

Hereditary non-polyposis colorectal cancer (HNPCC, aka Lynch syndrome) also involves variants in genes necessary for DNA repair

So, what about other hereditary cancer syndromes? The BRCA1 and BRCA2 proteins repair damage related to DNA breakage, are there other mechanisms that also lead to the accumulation of DNA damage? In fact, damage can also occur when the DNA is not faithfully copied during normal cell division, which requires a different and elegant corrective mechanism known as *DNA mismatch repair (MMR)*.[11] The failure of this process is associated with Lynch syndrome. As previously mentioned, each round of division in actively dividing cells, generates two daughter cells which contain perfect copies of the parent's DNA. DNA replication takes the somatic cell's original DNA, separates the two entwined strands from each other, and uses each as a template for two new strands of DNA. Each newly synthesized and identical strand is then allocated to one of two daughter cells (Fig. 4). Whenever the template strand carries an A, a T is added to the new strand and vice versa. Similarly, a G is added to complement a C, or vice versa. During the process of DNA replication, a proofreading function, DNA mismatch repair, corrects mismatched nucleotide insertions (such as an A with a C) (Fig. 5), but the process is not perfect and the average rate of mismatch errors is 1 in 100 million per DNA base pair per replication cycle. However, when assessing cancer risk, these rare errors lie at the core of a probability game. Most (99%) are eventually corrected by mismatch repair, so the net rate of new mutations arising from a failure of mismatch repair per replication cycle is ~ 1 in 10 billion per base pair—substantially less than one new mutation per cell division, but there are more considerations when assessing the risk of developing cancer.[12] First, there are trillions of cellular DNA replications

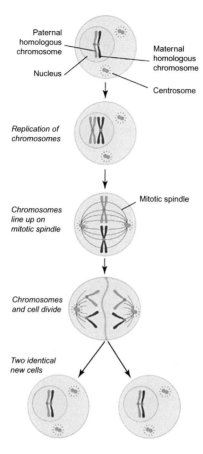

FIG. 4 Process of DNA replication of chromosomes in a somatic cell, followed by mitosis and separation of unreplicated chromosomes to two daughter cells. Errors during DNA replication result in DNA base mismatches requiring repair of the DNA sequence. The mismatch repair process facilitates maintaining the genetic identity of parent and daughter cells. *(From Korf BR, Sathienkijkanchai A. Introduction to human genetics. In: Clinical and Translational Science; 2009:265–287.)*

over the course of a person's life. Secondly, while the rates of cell division (and thus the absolute likelihood of a replication error) vary widely among individual cell types, an error that results in the impairment of cellular growth regulation processes, only needs to arise once to start down the path towards cancer. If the error is passed to its daughter cells, and those cells pass it to their daughters, cancer can develop quickly. With 6 billion bases to replicate (3 billion from each parent), the absolute replication error frequency is low for each cell division, but like the BRCA pathogenic variants, the failure of a mismatch repair protein need only occur once to have catastrophic consequences.

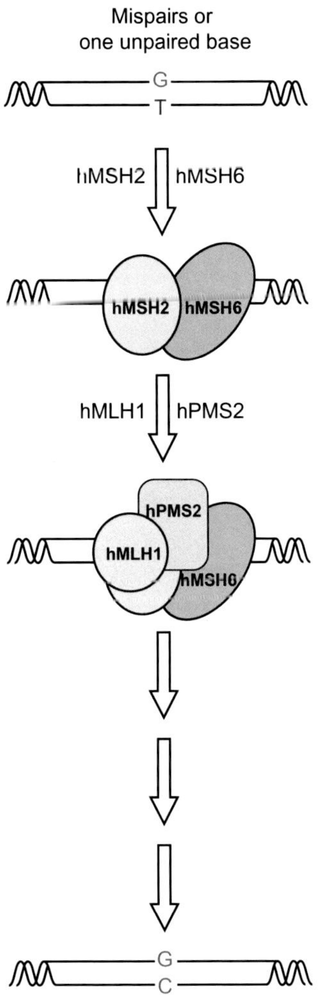

FIG. 5 Schematic representation of molecular events associated with DNA mismatch repair. Improper base pairing or the absence of a paired base (i.e., a single base deletion on one strand) between DNA strands is recognized by a pair of proteins: hMSH2 (human MSH2) and hMSH6, which in turn recruit other proteins, hPMS2 and hMLH1, that break the sugar-phosphate backbone of the DNA, remove bases surrounding the mispairing site, correct the error, and religate the DNA's sugar-phosphate backbone. A similar mechanism is employed for deletions of 2–4 bases on a single DNA strand. Some variants within the genes encoding these proteins: *MSH2, MSH6, MLH1,* and *PMS2* predispose a carrier to Lynch syndrome. If the second normal copy of the same gene is also damaged, then the cell loses its ability to repair DNA errors, resulting in the likelihood of cancer development. *(Modified from Kunkel TA. DNA mismatch repair: the intracacies of eukaryotic spell-checking. Curr Biol 1995;5:1091-94. https://doi.org/10.1016/S0960-9822(95)000218-1.)*

Mismatch repair involves a different brigade of proteins than those responsible for repairing double-stranded DNA breakages. The genes most commonly associated with mismatch repair proteins are: *MLH1*, *MSH2*, *MSH6*, and *PMS2*, all of which, interact to oversee the orderly restoration of the original DNA sequence, in the same way that *BRCA1*, *BRCA2*, and their protein partners, carry out repair of double-stranded DNA damage. The progression to cancer initiated by mismatch repair failure is depicted in Fig. 6.

What about the incomplete penetrance question? How does that factor in to hereditary syndromes related to DNA repair pathways? Remember the discussion of heterozygosity in Chapter 3? There are two copies of every gene, one on the chromosome inherited from the mother and one on the chromosome inherited from the father. If one gene in a cell is not functioning properly, the gene on the other chromosome can jump in and fill the gap. So, when an individual carries a pathogenic variant in one highly penetrant gene, the mismatch repair process can continue normally, because of the "normal" copy. However, if the functional copy is damaged at some point (due to environmental exposures or imperfect DNA copying), the cell loses its mismatch repair capability completely, and the result is an accumulation of variants from secondary genetic damage, and eventually loss of cellular growth control. Notably, in the rare case when an individual inherits a pathogenic variant that affects DNA repair (either mismatch repair (MMR) or DNA breakage repair (BRCA)) from both parents, cancer typically occurs in early in childhood.[13,14]

The importance of maintaining the integrity of the genome sequence is evident from the multiple mechanisms that have evolved to correct errors and restore the original DNA sequence, even though, overall, errors occur at a low rate. Pathogenic variants in genes involved in either of the two DNA repair mechanisms increase the lifetime *risk* for developing cancer. However, an adult who inherits a potentially pathogenic variant from one parent also, usually, inherits a normal and fully functional variant from the other.

A pathogenic variant increases a carrier's risk for HBOC and Lynch syndrome, but onset requires a second environmentally-induced event

Variants of DNA repair genes are known to increase the risk for several types of cancer, and many are associated with hereditary cancer syndromes, but inheriting a variant is not sufficient by itself to bring about cancer onset. As described above, this is largely due to having a second "normal" copy of the gene. When the second gene remains functional, carriers may never develop cancer, explaining the variable penetrance of many hereditary syndromes (note that incomplete penetrance mechanisms will vary based upon the cellular events that lead to disease). In this situation, the cell is heterozygous

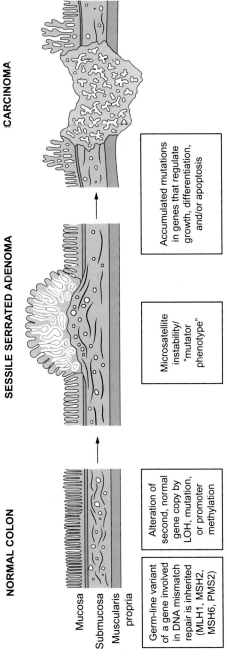

NORMAL COLON

SESSILE SERRATED ADENOMA

CARCINOMA

Mucosa
Submucosa
Muscularis
propria

Germ-line variant
of a gene involved
in DNA mismatch
repair is inherited
(MLH1, MSH2,
MSH6, PMS2)

Alteration of
second, normal
gene copy by
LOH, mutation,
or promoter
methylation

Microsatellite
instability/
"mutator
phenotype"

Accumulated mutations
in genes that regulate
growth, differentiation,
and/or apoptosis

FIG. 6 Graphical depiction of the developmental events that accompany the mutation of DNA mismatch repair genes. An individual inherits a pathogenic genetic variant but no effects are observed as long as the copy of the mismatch repair (MMR) gene inherited from the other parent performs normally. If the second copy of the MMR gene is damaged or inactivated, an adenoma forms and the cell's ability to repair DNA errors is impaired. In turn, this leads to further cellular degeneration as new and unrepaired mutations arise in the cell. Ultimately, the genetically damaged cell loses its ability to regulate cellular growth resulting in a carcinoma.

for the normal gene variant (e.g., *BRCA1 +/BRCA1 −*; *MLH1 +/MLH1 −*). However, the chances are relatively high that a mutational event will disrupt the normal gene copy sooner or later, that is, a cell in a carrier may eventually experience *loss of heterozygosity* (Fig. 7). If that happens, the cell no longer produces a normal mismatch repair protein and will lose its DNA repair capability. In those that inherit two functional gene copies, both copies need to be disrupted *in the same cell lineage* before loss of DNA repair capability occurs, that is, their cells are far less vulnerable to losing DNA repair function. When this occurs, the cancer syndrome is usually "sporadic," though, if it is induced by a highly hazardous environment that the entire family is exposed to, it may appear to be hereditary.

Errors occur frequently enough in individual cells that without an intact repair mechanism, it is highly likely that a cancer-causing gene variant will arise during the normal course of cell growth and division. Once a cell becomes cancerous, its uncontrolled growth results in rapid proliferation of its descendant cells. Currently there is no medical strategy for preventing cancer in individuals with HBOC except for prophylactic surgery and with Lynch syndrome except intensive surveillance for and removal of pre-cancerous lesions.

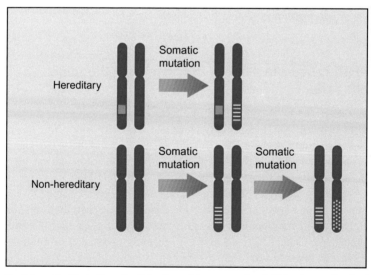

FIG. 7 Schematic illustration of cellular loss of heterozygosity (LOH) of genes involved with DNA repair as described in text. One homologous chromosome carries a pathogenic variant and the other carries a normal variant. If the remaining normal variant is damaged, the cell loses its capability for DNA repair. LOH is responsible for the onset of HBOC and Lynch syndrome. *Reprinted with permission from Elsevier. Jozwiak J, Jozwiak S, Wlodarski P. Possible mechanisms of disease development in tuberous sclerosis. Lancet Oncol. 2008;9:73–79. https://doi.org/10.1016/S1470-2045(07)70411-4.*

The requirement for a second hit or loss of heterozygosity explains why some individuals manage to escape without developing cancer even when they carry one of the high penetrance pathogenic variants we've described. The pathway to cancer is enabled by random mutational events, but the *BRCA1, BRCA2*, and mismatch repair variants do not cause cancer—they only make it more likely that it will occur, since the cell loses its ability to repair damage to cellular DNA. One question that follows is: why do some variants affecting the structure of repair genes exert a lower penetrance than others on cancer risk? The answer lies in the protein structure. Not all structural alterations impair a protein's function greatly enough to elevate the risk of cancer onset. In other words, the affected gene can create a *variant* protein that sometimes retains some or all of its DNA repair capacity. If DNA repair function is normal or near-normal, the variant is nonpathogenic (benign). If it is not-normal but still retains some function, it may increase the risk of developing cancer, but at a much lower rate than the ones we've been discussing. For instance, over 1000 *BRCA1* variants have been catalogued, but most are classified as VUSs. Ultimately, the pathogenicity of any variant must be confirmed by its clinical effects. For newly identified variants, family health history in conjunction with cosegregation analyses (as described in Chapter 4) are necessary for differentiating pathogenic and benign variants. The implications of family health history and its relationship to genetic variation and cancer risk will be explored later in the book, and it necessitates important considerations both for conducting medical practice and avoiding liabilities.

Different cancers in a family can arise from the same genetic variant

As our general understanding about the roles of individual genes and cellular processes evolves, it is increasingly clear that a predisposition to cancer may manifest in a variety of ways. For example, an individual with a sebaceous gland tumor, may be a warning sign to ask about relatives with colorectal cancer. What does a skin cancer and colon cancer have in common? How can they be related? Well, they are both epithelial cells and they rely on the same DNA repair mechanism, so on a cellular level, they are connected. In fact, their relationship is clearly seen in families with Lynch Syndrome. So clear that the National Comprehensive Cancer Network (NCCN) recently updated their guidelines to add sebaceous gland tumors. Perhaps the consistent principle concerning tumor suppressor and growth regulatory genes is that their impact on patient risk for cancer depends upon which cell types normally require them for growth regulation. The spatial and temporal pattern must be measured (and has been) for each gene in individual cell types to explain what may seem like unrelated cancer types. Even then, gene activity may vary within cell types, even in a single tissue, but this level of specificity lies beyond the level of detection for technologies approved and used in medical practice at this time.

How does a variant affecting DNA repair influence the type of cancer that develops? Where and when a cancer develops depends entirely upon where and when the remaining normal gene copy is mutated. The DNA repair genes are transcribed in a wide range of cell types, carrying out their repair functions throughout the body. Nevertheless, the number and rate of dividing cells varies widely from one cell type to another and so, some tissues are more prone to developing cancer than others. It follows that it is possible for different members of the same family to develop different cancer types, even when they carry the same genetic variant. In Lynch syndrome families, the incidence of several less common cancers is considerably more frequently than in the general population (Table 1). A similarly diverse pattern exists for HBOC syndromes. While the incidence of breast cancer (female and male) and ovarian cancer is quite high, so are other less common cancers (Table 2).[15] While BRCA1 and BRCA2 impact different points in the same DNA repair process, their effect on risk for other cancers is not equivalent. For example, prostate, cervical, and pancreatic cancer incidence is significantly higher among *BRCA2* variant carriers, while melanoma incidence is greater among *BRCA1* carriers. The takeaway is that when collecting family health history, it is critical for the provider to be aware that relatives carrying the same predisposing cancer variant can develop different cancer types, and in some cases, the cancers are of a relatively rare form that would normally be characterized as "sporadic." By corollary, cancer patterns (type and age of onset) in a family can offer clues to the underlying genetic basis of the syndrome. In fact, this is far more helpful than trying to map common cancers onto common cancer-causing genes. When the two are overlaid the emergent pattern represents a hodgepodge of genes and cancers in a many to many relationship; however, a collective family health history can, in some cases, be ascribed to a specific genetic variant (Fig. 8).

Early age of onset, severity, and recurrence are evidence of a hereditary cancer syndrome

When a parent or sibling develops cancer before the age of 50, guidelines recommend preventive action *even if no other family information exists*. If a parent develops colorectal cancer at age 38, siblings and children should begin colonoscopy screening at 28 (10 years prior to age of onset). For instance, the mother of President Bush's White House press secretary, Tony Snow, was afflicted by colorectal cancer at age 38 and died soon after. Mr. Snow developed colorectal cancer in his late 40s and passed away after a recurrence a few years later. Both the early age of onset in Mr. Snow and his mother, and the recurrence of the cancer, are "red flags" that a hereditary cancer is involved. There is no public information about the medical details which surrounded Mr. Snow's case, except that his mother's early cancer was sufficient to initiate early colonoscopy screening. A study of cases like Mr. Snow's found that ~20% of individuals had polyps in the colon before age 50.[16]

TABLE 2 Observed vs. expected incidence of several selected cancers in 1072 total carriers (male and female) of *ERCA1* or *BRCA2* variants.[15]

Cancer type	Gene	Observed No. of cases	Expected No. of cases	SIR	95% CI	P value
Breast (♀)	BRCA1	345	9.35	36.9	33.11–41.01	≤.001*
	BRCA2	246	8.88	27.69	24.34–31.74	≤.001*
Cervical	BRCA1	2	1.7	1.18	0.13–4.24	.99
	BRCA2	6	1.36	4.41	1.61–9.60	.006
Colorectal	BRCA1	6	1.58	1.59	0.58–3.44	.37
	BRCA2	2	0.53	0.53	0.06–1.91	.54
Hodgkin lymphoma	BRCA1	3	0.79	3.79	0.76–11.07	.09
	BRCA2	0	0.63	0	0–5.79	.93
NonHodgkin lymphoma	BRCA1	0	2.11	0	0–.74	.24
	BRCA2	1	1.98	0.51	0.01–2.81	.83
Leukemia	BRCA1	5	1.69	2.95	0.95–6.89	.06
	BRCA2	3	1.49	2.01	0.41–5.87	.38

Ovarian	BRCA1	178	1.28	139.12	119.43–161.12	≤.001*
	BRCA2	87	1.61	74.93	60.01–92.42	≤.001*
Pancreatic	BRCA1	4	0.85	4.73	1.27–12.11	.02
	BRCA2	19	0.87	21.75	13.09–33.96	≤.001*
Prostate	BRCA1	3	1.79	3.81	0.77–11.13	.09
	BRCA2	7	4.89	4.89	1.96–10.08	.002*
Skin (melanoma)	BRCA1	9	2.72	3.31	1.51–6.29	.004
	BRCA2	2	2.46	0.82	0.91–2.94	.89
Uterine	BRCA1	4	2.87	1.39	0.38–3.57	.65
	BRCA2	3	2.64	1.14	0.23–3.33	.98

Standardized incidence ratio (SIR) reports the actual number of cases recorded relative to the expected number based on population measures of incidence in a comparable population cohort. CI, confidence interval. Level of significance ≤.0025 indicated by (*) with calculated P value.

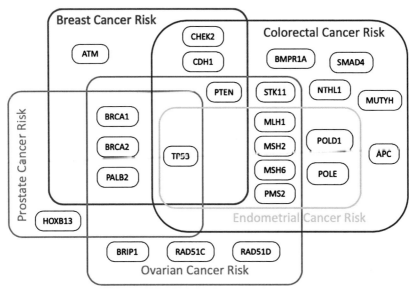

FIG. 8 Relationship between specific cancer-associated genes *(gene names given in black font)* and several cancer types. The list does not include several uncommon cancer types which occur at a relatively high frequency in carriers of pathogenic variants for some of the genes shown here.

Cancer occurrence in second degree relatives may be necessary to diagnose HBOC or Lynch syndromes

Given the small size of many modern families and the incomplete penetrance of hereditary syndromes, there may be no affected first-degree relatives in a family, even when they harbor a pathogenic variant. In addition, the total number of affected individuals will be small. This complicates diagnosis. A very good family physician once told me that if two or three first degree relatives were diagnosed with the same cancer, it was a warning sign. Unfortunately, that criterion excludes the majority of hereditary cancer cases. Considering typical family size, probabilities of Mendelian inheritance, and incomplete penetrance it is difficult to find enough affected family members, or in some cases any affected family member, to diagnose a hereditary cancer syndrome when relying entirely on first-degree relatives. The value of gathering a complete family health history is not simply about polling more family members, but that it can make an important contribution to understanding whether a cancer has hereditary underpinnings. This also explains why the current "gold standard" for family history collection by genetic counselors is a three-generation pedigree.

A common situation that highlights the potential importance of second-degree relatives for making a family diagnosis, and also happens to address misconceptions about paternal inheritance of HBOC variants, is presented here. A 43-year-old woman and mother of two sons thought that she had no family

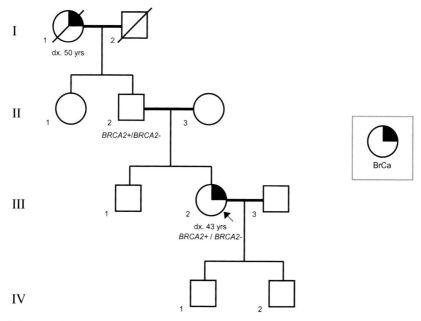

FIG. 9 Pedigree of family diagnosed with HBOC syndrome via family health history and positive *BRCA2* test result.

history of cancer. She was thus surprised when she developed an aggressive "triple negative" breast cancer (Fig. 9; III-2). Her physician encouraged her to seek more information from her relatives, which she did. Her query revealed that her deceased *paternal* grandmother (I-1) had developed breast cancer and passed away in her late 40s. Since her father was alive (II-2), he underwent genetic testing, and was found to have a pathogenic *BRCA2* variant, which she inherited. He was unaware that his *BRCA2* variant elevated his own risk for aggressive prostate cancer, though he was still healthy. In this case, the syndrome seemed to have "skipped" a generation, though in fact it was due to the protection afforded him by his normal *BRCA2* gene, on his other chromosome (the one from his father). Only her grandparents' health information suggested she had a high risk for cancer, before her actual encounter with breast cancer. This case also highlights the potential benefit of testing her sons (IV-1,2), when they reach adulthood, not only to determine whether one or both carry the hereditary syndrome and thus are at increased risk for cancer themselves, but also for the role it may eventually play in their offspring.

A similar misconception about gender and family health history applies to a second case (Fig. 10). In this instance, a 45-year-old father of a 12-year-old girl expressed concern to his physician about reaching the age at which his mother and his maternal aunt had been diagnosed with breast cancer, ages 42 and 46, respectively (II-1,2). There was little other information about the

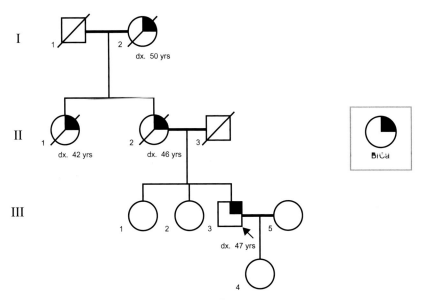

FIG. 10 Pedigree of family diagnosed with HBOC syndrome based on family health history with negative genetic test result. See discussion in text.

family available, because his maternal grandmother (I-2) had been adopted. His wife's family had no known family history of cancer. A few years later, he noticed a persistent breast discharge. He was diagnosed with male breast cancer and the conversation about family history was resurrected. Ultimately, he underwent genetic testing with a 32 gene hereditary breast cancer panel, but no pathogenic variants were found. He later learned that his paternal grandmother had developed breast cancer at age 50. His family history is very concerning, both for the risk it conferred on him, which culminated in his developing breast cancer, but also its implications for his daughter's risk. The negative breast cancer panel does not negate that risk, it simply prevents him from putting the name of a hereditary syndrome on his list of diagnoses; and prevents his daughter from being able to definitively determine whether she carries the same level of risk or not. In the absence of that information, she should be treated as high risk for breast cancer, based on her family health history alone. While this case did involve a first degree relative, the mother, there was considerable information also present in the second-degree relatives (maternal grandmother and maternal aunt) that alone would have warranted genetic testing. Recent studies point out the critical importance of cancers in second degree relatives, since moderately penetrant pathogenic variants may not be conclusively connected to familial cancers in nuclear families comprised of first-degree relatives.[17]

Diagnosing a hereditary condition in a family sometimes requires simultaneous consideration of several factors

From the cases presented above and the earlier discussions in this chapter, several specific principles emerge:

(1) Hereditary cancer syndromes can be responsible for cancer in a variety of cell types. For example, *BRCA1* and *BRCA2* pathogenic variants raise the risk of breast cancer in both genders, gender-specific cancers, and pancreatic cancer.

(2) Many cancer promoting variants have moderate pathogenicity and incomplete penetrance, that is, some carriers may never develop cancer, but still pass it along to their offspring (along with the elevated cancer risk).

(3) Mendelian inheritance principles apply to both genders, but the probability of disease onset depends upon cellular events that vary from one individual to another.

While each of the factors noted is straightforward, analyzing hereditary situations can become complicated, if they are not considered individually. An example of this complexity is presented in the following case. A 32-year-old woman (Fig. 11; III-1) knew that neither of her 60-plus-year old parents had a history of cancer. When she developed bloody stools, she discussed it with her mother, who reassured her, but encouraged her to mention it to her gynecologist. With prompting from the gynecologist, the woman remembered that her paternal grandfather had been diagnosed with testicular cancer at the age of 39 (I-2) and passed away at 42, and that her paternal aunt had been diagnosed with colorectal cancer at the age of 44 (II-4), but was still alive. In addition, she later found out her father had adenomatous polyps on his colonoscopies, but no cancer (II-3). He was aware of the potential importance of his own family health history, but was uncertain whether it was relevant for his children. Given

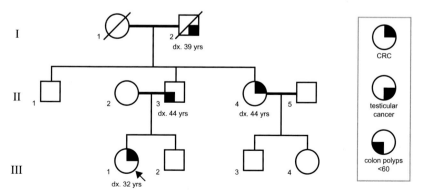

FIG. 11 Pedigree of family diagnosed with Lynch syndrome based on family health history with negative test result. See discussion in text.

the rectal bleeding, she underwent a colonoscopy, which detected a tumor in the sigmoid colon. Microsatellite instability (MSI) testing of the tumor identified a high level of genomic instability, a strong indication (but not diagnostic) of Lynch syndrome. Immunohistochemical staining (IHC) failed to detect MSH6, a mismatch repair protein in the sample. This led to genetic testing for both *MSH2* and *MSH6*, which was negative for a pathogenic variant. Despite the negative genetic test, her oncologist diagnosed her with Lynch syndrome based on her family history, her early onset disease, and her positive MSI test. Her colon cancer was treated and she began screening for colon cancer recurrence as well as other Lynch related cancers. Her relatives were also started on an intensified screening program. Notably, her two cousins (the daughters of the aunt with colon cancer) had already started screening colonoscopies at 34 based on current guidelines. If a pathogenic variant had been identified, it would have permitted "cascade" testing of other family members to determine, which ones had inherited the variant and its predisposition for cancer. As it stands now, they will all need to be managed based upon the elevated risk conferred by their family history, until a time when the etiology is identified and they can be tested. Given how little we know about genomics, for the foreseeable future, family health history will be the guiding source of information for advising and managing cancer risk in when genetic tests are "negative."

This family history brings up all three of the outlined concepts. First, the importance of collecting a family health history that includes second-degree relatives is evident, since, as in this case, it is possible that first-degree relatives with a pathogenic variant never develop cancer. The two second degree relatives (her father's two first degree relatives) with cancer at an early age offered important clues about her own risk. In addition, families with few or no siblings, or a lost parent, can be difficult to interpret in the absence of an extended family history. Second, when a variant is moderately penetrant, many carriers will not develop cancer, even though their overall cancer risk is elevated compared to the general population. Here again, more distant relatives (including third degree relatives such as cousins, great uncles and great aunts, etc.) can provide important information about a family's risk. Third, the paternal grandfather's testicular cancer is confusing. Why would the paternal grandfather's testicular cancer be included as part of her risk assessment? There are no studies conclusively linking testicular cancer to Lynch syndrome, but the rarity of testicular cancer makes it difficult to achieve the necessary statistical power to establish a connection. There are, however, case reports of testicular cancer in Lynch syndrome families,[18] leaving open the possibility of a functional connection. On its own the testicular cancer evidence is too weak to establish a diagnosis, but in conjunction with the other findings, provides supportive evidence.

What about her negative genetic test result? What does that mean for her diagnosis and for her family members? If Lynch syndrome is hereditary, why didn't she have a pathogenic variant on her genetic test? First, she was only tested for two of the mismatch repair genes, *MSH2* and *MSH6*. Second, most variants in the DNA repair genes have not been properly evaluated for pathogenicity (as discussed in Chapter 4) and thus remain classified as variants of undetermined significance. Further, it is likely that there are other genes involved in these cellular pathways that have yet to be discovered. Most are discovered either accidentally or after exhaustive efforts analyzing the genome. The lack of knowledge about most variants, even those that are plausibly pathogenic by virtue of being located in genes that regulate cell growth, profoundly limits the value of a "negative" genetic test. For now, family phenotype (e.g., polyps, tumor formation) will need to serve as a proxy for genotype.

In most, 95%, of colon cancers with a cellular loss of MSH6 protein (as seen in the case above), genetic testing identifies a pathogenic variant, but a conclusive genetic test result is not a requirement for a diagnosis of Lynch syndrome. In fact, no variant is found in ~7%, and a VUS in ~6%, of Lynch syndrome families. Recent data suggest that 80% of colon tumors with high microsatellite instability and negative genetic testing from blood or saliva samples (i.e., normal cells), have pathogenic variants in their cancer cells that are diagnostic for Lynch syndrome.[19–22] These usually present as sporadic cancers, but that is not the most likely explanation in this case.

Cancer occurrence sometimes involves non-genetic cellular mechanisms

The cancer twin study described in Chapter 1 found that cancer risk was attributable to a combination of genetic and environmental factors. The exact factors were unknown, but it was clear that genetics was not the sole driver of risk. Explanations for the impact of environment on disease risk include environmentally induced DNA damage, discussed earlier in this chapter, and possibly *epigenetic modification* of genes involved in DNA repair. There are several epigenetic mechanisms known, though the significance of these has not been established. One mechanism is *methylation*, a chemical modification of DNA that reduces a gene's activity and thus the amount of protein produced by the gene, without affecting the protein's sequence.[23] Translating to hereditary cancers, if the activity of a DNA repair gene is epigenetically suppressed, the amount of repair protein produced could be severely reduced, leading to the same outcome as when the repair protein is non-functional—accumulation of secondary genetic damage.[24] Little is known about this process and the cellular and/or environmental triggers for methylation and other epigenetic mechanisms associated with cancers have not been established, and thus cannot yet be intervened on.

Other hereditary cancers result from genetic variants disrupting other cellular mechanisms

As we've described in many cases hereditary cancers arise from variants that directly affect the repair of DNA damaged in the normal course of life. In these syndromes penetrance is generally high. Another genetic mechanism for hereditary cancers is related to variants that affect tumor suppressor proteins, which often have near complete penetrance. For example, familial adenomatous polyposis cancer and Lynch syndrome, both hereditary colon cancer syndromes, have distinct clinical and genetic findings. Familial adenomatous polyposis manifests with hundreds to thousands of polyps all along the GI tract, starting as early as adolescence,[25] while Lynch presents with single tumors that can appear in the third or fourth decade that frequently recurs. Familial adenomatous polyposis's phenotype is driven by the APC protein's role in cell function, initiation of programmed cell death (known as *apoptosis*). The cellular processes, which are not yet well understood, result in tumor suppression by activating a tumor cell's internal mechanism for apoptosis, thereby forcing it to kill itself.[26] In the traditional form of familial adenomatous polyposis, damage of just one gene copy, is enough to cause most of the proteins to be defective. The consequence is polyps—and lots of them. In this situation, no "second hit" is necessary. It's a completely penetrant, autosomal dominant variant just as Gregor Mendel described it. There are some milder "attenuated" variants, in which the protein has some function, leading to fewer polyps and a less dramatic disease course, though it is still associated with a large number of polyps and an ~30% lifetime risk of colorectal cancer. Is there a "second hit" impact in attenuated familial adenomatous polyposis? Right now, we don't know the answer to that question.

Another cellular mechanism recently brought to the forefront by Federal Drug Administration approval for DTC testing is the *MUTYH* gene, which is associated with a rare hereditary colorectal cancer. Pathogenic variants of this gene impair activity of the MUTYH enzyme, which reverses naturally occurring oxidative damage to cells during DNA replication. The FDA has approved direct to consumer testing of two founder variants in the gene. Inheriting one copy of either variant modestly increase colorectal cancer risk.[27] Note that as with other genetic tests, a negative test result conveys no information about an individual's actual cancer risk.

As illustrated by the distinction between Lynch syndrome and familial adenomatous polyposis presentation, the differential effect of genetic variants for many chronic diseases will result in a range of severity, age of onset, comorbidities, and recurrence risk. These differences are largely responsible for the different disease risk levels described in Chapter 2: population, familial, and hereditary. The more profound the cellular dysfunction the higher the risk associated with the variant and the more penetrant it will be. Given that a variant can only cause so much dysfunction before it kills its "host" (us), there is an upper limit on severity; but not a lower limit. In fact, the less severe the dysfunction the more likely the individual is to live to reproduce and perpetuate the variant. Thus, *the vast majority of families with a family history of cancer do not have a hereditary syndrome, but rather a familial level of risk.* They benefit from increased monitoring and risk management, but the strategies used are less aggressive than those with hereditary syndromes.

Pancreatic cancer involves a complex interaction of genetic variation and environment

When thinking about the relationship between genetics, family health history, and environmental risk factors, the epidemiology of pancreatic cancer offers a caveat about overgeneralizing gene-environment interactions. The lifetime incidence of pancreatic cancer in the United States is 1.5% (population level risk). An individual with a first degree relative with pancreatic cancer is four times more likely to develop pancreatic cancer than the general population (familial level risk). As the number of affected relatives increases, so does the risk; however, the actual level of risk is still quite low (i.e., 1.5% × 4-fold increase = 6% absolute risk).[28,29] Studies have also shown that tobacco smoking triples the risk of sporadic pancreatic cancer. So, what are the combined effects of a positive family health history and a personal history of smoking on pancreatic cancer risk? It has been difficult to determine. Environmental contributions to risk are confounded by the fact that pancreatic cancers occur in multiple different hereditary cancer syndromes, included HBOC and Lynch, as well as Familial Pancreatic Cancer and Familial Atypical Mole and Melanoma syndromes. In addition, most familial and sporadic cases of pancreatic cancer cannot be ascribed to a pathogenic variant. Though interestingly, pancreatic cancer in children of parents with pancreatic cancer arises an average of ten years earlier, suggesting that an unclear phenomenon known as genetic anticipation may influence age of onset.[28,29] With all these factors in play and the relative rarity of pancreatic cancer, it will take some time to fully understand the cellular mechanisms and their genetic and environmental drivers.

Genetic biomarkers could have utility for refining personal and familial cancer risk assessment

Both hereditary and familial levels of cancer risk elevate the urgency for enhanced risk management strategies, as well as the desire to have a reliable biomarker for the presence or absence of disease. An example, of a biomarker is the prostate specific antigen (PSA), which has been widely used to screen for prostate cancer, though its predictive and diagnostic value has been questioned frequently.[30] A genome wide association study (GWAS), conducted to see if genomic screening could enhance prostate cancer screening, identified 40 SNPs associated with elevated PSA levels. However, less than half had been previously associated with an elevated risk for prostate cancer.[31] PSA levels exemplify an *endophenotype*, that is, a trait or measurement tied more directly to genotype than a phenotype, such as cancer incidence. A wide range of clinical measurements can be viewed as endophenotypes that are at least partially dependent on a genotype. The study results illustrate how intermediate outcomes, like PSA, should not be used as outcomes in genomic association studies. If SNPs were associated with prostate cancer, rather than PSA, the test may become more useful.

Summary and conclusions

The relationship between family health history, genetic variation, environmental factors, and cancer incidence is complex and varied. Relatively few genetic variants directly cause adult cancer onset. Typically, variants alter a cell's function, increasing its (and our) vulnerability to disease. Frequently, unavoidable environmental exposures and random molecular events are the triggers that activate the pathway that leads to cancer onset, especially in those who have a genetic (family-based) vulnerability. The interface between the genome and the environment determines an individual's disease risk, and both need to be considered when contemplating a medical course of action. Patient management frequently requires more than simply leading an ideal lifestyle. In the optimal scenario we know the cause and impact of the cellular dysfunction and have an intervention that can restore normal function. Unfortunately, hereditary syndromes are complex and most of the pathogenic variants that drive them are unknown. In addition, we continue to generate and perpetuate new (de novo) variants that have never been seen before and family-specific variants that have not been discovered yet. Therefore, family health history is still the primary, and most effective, indicator of disease risk. When genetic information cannot be obtained or a genetic test yields uncertain (or no) findings, family health history alone provides a sufficient basis for recommending efficacious interventions. In those cases when we can identify a specific pathogenic variant in an affected patient, cascade testing facilitates risk assessment for family members.

Several points follow concerning the diagnosis and detection of familial cancer risks:

- Certain genetic variants alter cellular functions and their capacity to respond to environmental challenges, initiating a pathway that leads disease, though most variants exert no effect on function nor of course, on health.
- The pathway from genetic variant to cancer onset is usually (but not always) a multistep progression and monitoring milestones before cancer onset is useful.
- Some environmental risks are unavoidable and for individuals with a high inherent risk for certain diseases, intensive screening and monitoring is necessary.
- Even strongly pathogenic variants, like those for HBOC, Lynch syndrome, and other hereditary syndromes, are usually incompletely penetrant, making it difficult to identify at-risk families.
- Family health information from second degree (and sometimes more distant) relatives may help overcome challenges in diagnosis with small families and variable disease penetrance.
- The ability to detect and evaluate reliable biomarkers that precede and portend the likelihood of cancer onset is particularly crucial for families with an elevated cancer risk, even if a specific genetic variant cannot be identified.

All three authors contributed to the content of this chapter.

References

1. Chen S, Parmigiani G. Meta-analysis of BRCA1 and BRCA2 penetrance. *J Clin Oncol.* 2007;25:1329–1333.
2. Bunnell AE, Garby CA, Pearson EJ, Walker SA, Panos LE, Blum JL. The clinical utility of next generation sequencing results in a community-based hereditary cancer risk program. *J Genet Couns.* 2017;26:105–112. https://doi.org/10.1007/s10897-016-9985-2.
3. Kwong A, Shin VY, Au CH, et al. Detection of germline mutation in hereditary breast and/or ovarian cancers by next-generation sequencing on a four-gene panel. *J Mol Diagn.* 2016;18:580–594. https://doi.org/10.1016/j.jmoldx.2016.03.005.
4. Tung N, Lin NU, Kidd J, et al. Frequency of germline mutations in 25 cancer susceptibility genes in a sequential series of patients with breast cancer. *J Clin Oncol.* 2016;34:1460–1468. https://doi.org/10.1200/JCO.2015.65.0747.
5. Mannan AU, Singh J, Lakshmikeshava R, et al. Detection of high frequency of mutations in a breast and/or ovarian cancer cohort: implications of embracing a multi-gene panel in molecular diagnosis in India. *J Hum Genet.* 2016;61:515–522. https://doi.org/10.1038/jhg.2016.4.
6. Shirts BH, Casadei S, Jacobson AL, et al. Improving performance of multigene panels for genomic analysis of cancer predisposition. *Genet Med.* 2016;18:974–981. https://doi.org/10.1038/gim.2015.212.
7. Susswein LR, Marshall ML, Nusbaum R, et al. Pathogenic and likely pathogenic variant prevalence among the first 10,000 patients referred for next-generation cancer panel testing. *Genet Med.* 2016;18:823–832. https://doi.org/10.1038/gim.2015.166.
8. Lincoln SE, Kobayashi Y, Anderson MJ, et al. A systematic comparison of traditional and multigene panel testing for hereditary breast and ovarian cancer genes in more than 1000 patients. *J Mol Diagn.* 2015;17:533–544. https://doi.org/10.1016/j.jmoldx.2015.04.009.
9. Prakash R, Zhang Y, Feng W, Jasin M. Homologous recombination and human health: the roles of BRCA1, BRCA2, and associated proteins. *Cold Spring Harb Perspect Biol.* 2015;7:a016600. https://doi.org/10.1101/cshperspect.a016600.
10 Brianese RC, Nakamura KDM, Almeida FGDSR, et al. BRCA1 deficiency is a recurrent event in early-onset triple-negative breast cancer: a comprehensive analysis of germline mutations and somatic promoter methylation. *Breast Cancer Res Treat.* 2018;167:803–814. https://doi.org/10.1007/s10549-017-4552-6.
11. Kunkel TA. DNA mismatch repair: The intracacies of eukaryotic spell-checking. *Curr Biol.* 1995;5:1091–1094. https://doi.org/10.1016/S0960-9822(95)000218-1.
12. Li SKH, Martin A. Mismatch repair and colon cancer: mechanisms and therapies explored. *Trends Mol Med.* 2016;22:274–289. https://doi.org/10.1016/j.molmed.2016.02.003.
13. Nepal M, Che R, Zhang J, Ma C, Fei P. Fanconi anemia signaling and cancer. *Trends Cancer.* 2017;3:840–856. https://doi.org/10.1016/j.trecan.2017.10.005.
14. Levi Z, Kariv R, Barnes-Kedar I, et al. The gastrointestinal manifestation of constitutional mismatch repair deficiency syndrome: from a single adenoma to polyposis-like phenotype and early onset cancer. *Clin Genet.* 2015;88:474–478. https://doi.org/10.1111/cge.12518.
15. Mersch J, Jackson MA, Park M, et al. Cancers associated with BRCA1 and BRCA2 mutations other than breast and ovarian. *Cancer.* 2015;121:269–275. https://doi.org/10.1002/cncr.29041.
16. Syrigos KN, Charalampopoulos A, Ho JL, Zbar A, Murday VA, Leicester RJ. Colonoscopy in asymptomatic individuals with a family history of colorectal cancer. *Ann Surg Oncol.* 2002;9:439–443.
17. Solomon BL, Whitman T, Wood ME. Contribution of extended family history in assessment of risk for breast and colon cancer. *BMC Fam Pract.* 2016;17:126. https://doi.org/10.1186/s12875-016-0521-0.

18. Huang D, Matin SF, Lawrentschuk N, Roupret M. Systematic review: an update on the spectrum of urological malignancies in lynch syndrome. *Bladder Cancer*. 2018;4:261–268. https://doi.org/10.3233/BLC-180180.

19. Sourrouille I, Coulet F, Lefevre JH, et al. Somatic mosaicism and double somatic hits can lead to MSI colorectal tumors. *Fam Cancer*. 2013;12:27–33. https://doi.org/10.1007/s10689-012-9568-9.

20. Mensenkamp AR, Vogelaar IP, van Zelst-Stams WA, et al. Somatic mutations in MLH1 and MSH2 are a frequent cause of mismatch-repair deficiency in Lynch syndrome-like tumors. *Gastroenterology*. 2013;146:643–646.e8. https://doi.org/10.1053/j.gastro.2013.12.002.

21. Geurts-Giele WR, Leenen CH, Dubbink HJ, et al. Somatic aberrations of mismatch repair genes as a cause of microsatellite-unstable cancers. *J Pathol*. 2014;234:548–559. https://doi.org/10.1002/path.4419.

22. Haraldsdottir S, Hampel H, Tomsic J, et al. Colon and endometrial cancers with mismatch repair deficiency can arise from somatic, rather than germline, mutations. *Gastroenterology*. 2014;147:1308–1316.e1. https://doi.org/10.1053/j.gastro.2014.08.041.

23. Wu H, Zhang Y. Reversing DNA methylation: mechanisms, genomics, and biological functions. *Cell*. 2014;156:45–68. https://doi.org/10.1016/j.cell.2013.12.019.

24. Downs B, Wang SM. Epigenetic changes in BRCA1-mutated familial breast cancer. *Cancer Genet*. 2015;208:237–240. https://doi.org/10.1016/j.cancergen.2015.02.001.

25. Kerr SE, Thomas CB, Thibodeau SN, Ferber MJ, Halling KC. APC germline mutations in individuals being evaluated for familial adenomatous polyposis: a review of the Mayo Clinic experience with 1591 consecutive tests. *J Mol Diagn*. 2013;15:31–43. https://doi.org/10.1016/j.jmoldx.2012.07.005.

26. Fodde R. The APC gene in colorectal cancer. *Eur J Cancer*. 2002;38:867–871.

27. Mazzei F, Viel A, Bignami M. Role of MUTYH in human cancer. *Mutat Res*. 2013;743-744:33–43. https://doi.org/10.1016/j.mrfmmm.2013.03.003.

28. Matsubayashi H, Takaori K, Morizane C, et al. Familial pancreatic cancer: concept, management and issues. *World J Gastroenterol*. 2017;23:935–948. https://doi.org/10.3748/wjg.v23.i6.935.

29. Petersen GM. Familial pancreatic cancer. *Semin Oncol*. 2016;43:548–553. https://doi.org/10.1053/j.seminoncol.2016.09.002.

30. Hayes JH, Barry MJ. Screening for prostate cancer with the prostate-specific antigen test: a review of current evidence. *JAMA*. 2014;311:1143–1149. https://doi.org/10.1001/jama.2014.2085.

31. Hoffmann TJ, Passarelli MN, Graff RE, et al. Genome-wide association study of prostate-specific antigen levels identifies novel loci independent of prostate cancer. *Nat Commun*. 2017;8:14248. https://doi.org/10.1038/ncomms14248.

Chapter 6

Using family health history to identify and reduce modifiable disease risks

Vincent C. Henrich, Lori A. Orlando, and Brian H. Shirts

- Cardiovascular disease risk in the population is attributable to the independent contribution of genetic and environmental risk factors.
- The additive and interactive effects of genetic variants with each other and the environment provide a population measure of disease risk.
- The relationship between inherited variants and environmental risks depends on a specific condition and may require a specific treatment or intervention.
- The health effects of controllable environmental factors in individuals and families are not entirely understood.
- Using family health history to assess future health risk.
- Family health history *is* applied genomics.
- Summary and conclusions.

In the previous chapter, we explored the progression of events leading from genetic variant to cancer onset. As seen for at least some major hereditary cancer syndromes, unavoidable environmental exposures are responsible for the transition from risk to actual incidence. Even for other cancers, the effect of a genetic variant for cancer risk and severity is difficult to predict and likely subject to environmental modification. When the clinical presentation indicates an underlying genetic process, finding and confirming a genetic "cause" still is frequently futile. The question to be addressed here is: Do modifiable environmental factors influence a family-based disease risk? If so, how can these risks be surmised early and followed by interventions, screenings, and treatments that ameliorate an intrinsic risk? As will be seen for several chronic diseases and predisposing medical conditions, environmental risk factors can be controlled enough to reduce risk, and while a familial risk may not be eliminated, proactive and targeted management and monitoring can benefit potentially affected family members.

Managing Health in the Genomic Era. https://doi.org/10.1016/B978-0-12-816015-2.00006-7

The combined impact of a positive family history for venous thrombosis and oral contraceptive use on individual risk for thrombophilia has been established repeatedly in the medical literature and offers a straightforward example about how genetic and environmental risk factors can compound a health risk. A 2015 Swedish study[1] reported that the odds ratio of developing venous thrombosis for a woman with a positive family health history of venous thrombosis is 2.53 (95% CI: 2.23–2.87). Similarly, the odds ratio for women using oral contraceptives is 2.38 (2.09–2.71), and for those with both a family health history and oral contraceptive use, the odds ratio is 6.02 (5.02–7.22). Even higher odds ratios have been reported when carriers of specific thrombophilic gene variants were tested. However, family health history may be the more reliant measure of this gene-environment interaction because the penetrance of variants such as Factor V Leiden for thrombosis is low and because a substantial fraction of familial thromboses cannot be ascribed to any of the known genetic variants.

Cardiovascular disease risk in the population is attributable to the independent contribution of genetic and environmental risk factors

The relative influence of inherited and environmental risk on disease incidence has been debated continuously over the years, as acknowledged in Chapter 1. As more genetic variants have been implicated in disease risk, studies have attempted to address more directly how both influence disease incidence. Cardiovascular disease is the world's most common chronic condition. As will be explored in this chapter, environmental risks clearly increase the likelihood of disease in a family with a history of specific cardiovascular ailments. However, the relationship between genetic predisposition and environment is distinct from the two-step process described for some of the hereditary cancers discussed in the previous chapter. For cardiovascular disease, the risk-elevating impact of environment is quantitative. For these diseases, one or more environmental risk factors interact with a predisposing genotype to increase the probability of an adverse health event. From the patient care perspective, altering environmental and lifestyle factors leads to a concomitant reduction in the risk of an adverse change in health status. In other words, these environmental risk factors can be controlled more readily than those that might trigger a cancer in a genetically vulnerable cell.

This quantitative relationship was tested in a study of over 300,000 subjects by parsing the additive and interactive effects of genetic variation and environment.[2] Subjects were tested for several known single nucleotide polymorphisms (SNPs), identified by a broad range of studies covering a variety of cardiovascular diseases. Based on their SNP genotypes, each subject was assigned a polygenic score and classified at low, intermediate, or high genetic risk for cardiovascular disease. The same subjects also scored themselves as ideal, average, or poor in regards to their personal lifestyle choices. When the entire aggregate was analyzed, cardiovascular disease risk was found to depend on the additive and independent influence of genetics and environment, for all genetic risk and lifestyle categories. The study population was broad and diverse, capturing a large number of different genetic and environmental risks factors at play, but there were no interactions that affected the accuracy of the model. Nevertheless, there is little power to identify specific gene-environment interactions in such a large-scale study, as will be further explained below. The overall message is that for the risk of cardiovascular disease, one's lifestyle choices affects one's disease risk regardless of inherent risk level. Several ancillary points are drawn from a closer inspection of the data. Even in the low genetic risk group, poor lifestyle substantially increased the risk of coronary artery disease, hypertension, and type 2 diabetes. Just as importantly, someone with a low genetic risk cannot expect to be spared the consequences of poor lifestyle habits (Fig. 1)[2]. On the other hand, a healthy personal lifestyle offered a significant benefit to subjects in all genetic risk groups. Both the benefit and risk of lifestyle choices were especially pronounced for individuals in the high genetic risk group. For type 2 diabetes, stroke, and hypertension, poor lifestyle is the primary contributor for converting a genetically-based risk into a change in health status, though even an ideal lifestyle did not fully eliminate a genetically-based risk for coronary artery disease or atrial fibrillation. The variable effect of intervention for different diseases may prove to have implications for refining health management; some cardiovascular and metabolic conditions may prove to be more affected by intervention than others. The beneficial value of intervention was especially notable for type 2 diabetes and hypertension, where a high genetic risk coupled with poor lifestyle substantially increased the likelihood of disease onset. On the other hand, an ideal lifestyle almost completely offset these risks, especially in the high genetic risk cohort.

The net effects across all groups may cancel or obscure specific interactions, so the practitioner may need to rely on a more precise evaluation of the patient's health status. When thinking about personal disease risk and the polygenic risk score employed in the study described above, an obvious question is: which individuals are most likely to carry the *same* genetic variants (i.e., structurally identical in terms of their predicted effect on health status) that led to a high genetic risk score in the study subjects? The obvious answer is: the subject's first degree relatives. The information drawn from second degree relatives is also potentially useful as they share a smaller, but substantial, subset of predisposing variants. If a half-sibling, aunt, uncle, or grandparent developed a familial condition, it is quite plausible that they harbor one or more shared variants that exert a major effect on disease risk. Even without knowing the genetic underpinning, similarities in clinical presentation are expected, and similar interventions are more likely to be helpful in a well-tailored management regimen.

In summary, modifying an environmental risk may yield a disproportionate benefit for a patient in a moderate to high disease risk category who has knowledge of an affected relative, particularly if exacerbating environmental factors

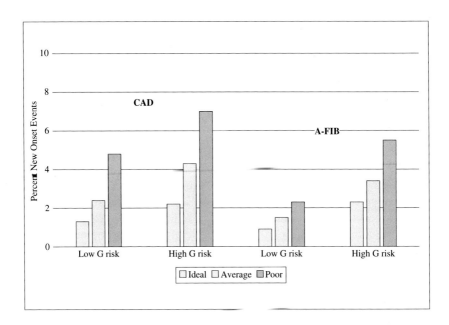

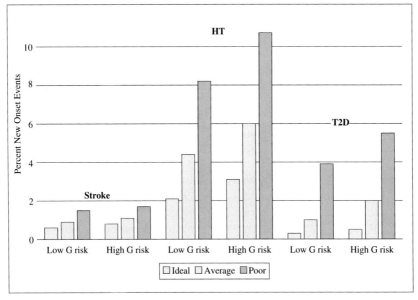

FIG. 1 Percent new-onset events of several cardiovascular disease and condition onsets among subjects with previously measured low and high genetic (G) risk scores and ideal (I), average (A), and poor (P) lifestyle factors as reported by Said et al.[2] The entire study involved ~325,000 subjects. Of these, ~67,000 were scored as low G risk (low genetic risk) and ~65,000 were scored in the high G risk category. Intermediate genetic risk categories are not shown here. Abbreviations: *CAD*, coronary artery disease; *A-FIB*, atrial fibrillation; *HT*, hypertension; *T2D*, type 2 diabetes.

are shared. If known interventions benefited relatives, they are also more likely to benefit a patient. This benefit, both medically and economically, is based on the same rationale applied for recommending early colonoscopies (i.e., before age 50) to family members related to an individual diagnosed with colon cancer before age 60.

The additive and interactive effects of genetic variants with each other and the environment provide a population measure of disease risk

Genetic modifier variants are a special case of what is often termed a "polygenic" effect. A number of familial adult onset diseases are described as "polygenic" infering that genetic risk is additive and incremental. Polygenic risk scores incorporate many risk variants and do not rely on individual genetic influences. This view implies that the individual variants do not increase risk enough to be designated as pathogenic by themselves. Current implementations of polygenic risk scores do *not* incorporate data for uncommon and family-specific pathogenic variants or known highly penetrant monogenic mutations, such as *BRCA1/2* variants. Instead, they utilize SNPs and indels (insertions or deletions) discovered primarily through genome wide association studies. In fact, every ongoing biological process occurring in a human is polygenic, that is, the process involves a number of different proteins, each encoded by a gene and interacting physically and functionally in a myriad of ways. When viewed from a population standpoint, there are an enormous number of genetic variants that could conceivably affect and even impair any number of vital functions—cell growth, cardiovascular performance, metabolism, immune response, neurophysiology and cognitive function, to name a few. Molecular pathways are designed to maintain homeostasis within an organism in an ever-changing environment. Environmental interplay that disrupts genetically defined molecular pathways for maintaining homeostasis leads to disease. If one takes a population standpoint about disease risk, it is apparent that there are many possible genetic breakdowns and that any one of them or combinations of them might lead to a given health event. These genetic combinations are most likely to recur among close relatives of a patient.

Alzheimer's disease provides an example of the distinction between population and individual genetic risk. The $APOE^{e4}$ variant is common and raises the risk of late onset Alzheimer's disease (as discussed in Chapter 3), but a significant proportion of familial Alzheimer's disease cases do not involve this variant. Other genetic variants implicated in familial Alzheimer's disease display a high variability in penetrance and age of onset.[3] For those Alzheimer's cases that do not involve $APOE^{e4}$, the extent to which polygenic action in an individual and/or environmental interaction are responsible for incidence is also unknown. Polygenic risk scores for patients with a positive family health history for Alzheimer's have been introduced[4] and even more genes have been reported with variants that heighten Alzheimer's disease risk though these have not been subjected to rigorous analysis in the general population.[5,6] The Alzheimer's Association has cautioned that the utility of genetic tests for Alzheimer's disease risk still has not been validated sufficiently to recommend their use. Another study recently concluded that examining the health history of ancestors such as great aunts and uncles may be especially useful for establishing a genetically based familial risk even if there is little possibility of pinpointing a specific genetic variant, an acknowledgment of the continuing value of family health history.[7] From a practice standpoint, it is imperative to make a careful evaluation of the presentation (phenotype), and to be particularly mindful that some family members are likely to share common presentations given their similar genetic makeup and shared environmental exposures. For Alzheimer's disease, a familial variant could influence important parameters including average age of onset, the rate of disease progression, and the severity of Alzheimer's disease.

The relationship between inherited variants and environmental risks depends on a specific condition and may require a specific treatment or intervention

Population studies of polygenic risk scores to measure genetic risk, and self-reported lifestyle to measure environmental risk, provide strong statistical evidence that both genetics and environment contribute to the risk of disease onset. However, the extent and nature of the wide variety of gene-environment relationships at the individual level are not well represented by statistical population measurements. It is critical to keep this in mind when assessing risk in individuals and their families.

Cholesterol regulation, and genetic variations of genes controlling cholesterol regulation, are known to exert a significant effect on a patient's personal risk for several diseases, as noted earlier for the $apoE^{\varepsilon 4}$ variant. Lipid regulation is a complex process which is partially understood and involves a host of lipid carrying proteins, lipoprotein receptors, metabolizing enzymes, and transport proteins. For the provider, the measurement of total cholesterol levels, high density lipoprotein-cholesterol, low density lipoprotein-cholesterol, and triglycerides offers a reasonable estimate of a patient's health status risk and is the cornerstone for making recommendations about statins and/or other cholesterol reducing drugs.[8] These test results represent the cumulative effect of both environment and inherited risk, and identifying familial patterns can be essential to determining an effective management strategy.

It is critical to recognize that managing a familial risk may depend upon specific and deliberate interventions to maintain good health status and avoid an adverse health event. Following an idealized lifestyle, by itself, may not be enough to forestall disease progression and family health history may offer key points of information. An instance in which the proactive use of family health history proved important involved a 57-year-old white male who learned that

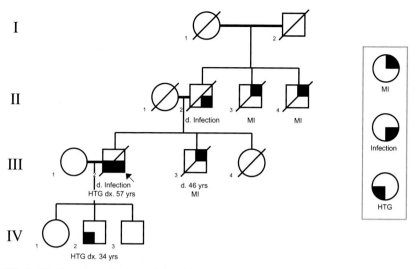

FIG. 2 Family pedigree for hypertriglyceridemia (HTG) and possible related health events.

his 46-year-old brother had been found deceased in a hotel room during a business trip. The younger brother, a moderate drinker and a heavy smoker, had suffered a sudden and catastrophic heart attack. The surviving brother was a pipe smoker, an infrequent alcohol consumer, did not exercise, and had a BMI over 30. The brothers' father had died from complications of an abscessed tooth at the age of 74 and their two paternal uncles had died of heart attacks in their late 50s and early 60s, respectively (Fig. 2).

The surviving brother (III-2) consulted with his physician and an initial screening performed to evaluate his cholesterol levels. It indicated that his high density lipoprotein levels were 50 mg/dL, low density lipoprotein levels were 160 mg/dL, total cholesterol level was 295 mg/dL, and his triglyceride level was 550 mg/dL. The physician recommended immediate enrollment in a weight loss-diet management program, strictly limiting caloric and carbohydrate intake, and initiating an exercise program. Over the course of 60 days, the patient lost 20 lbs. with improvement of his low density lipoprotein level, but not his high density lipoprotein or triglyceride levels. He continued to lose weight, but his triglyceride levels continued to increase, reaching ≥ 700 mg/dL. After 6 months on the program, the patient had brought his BMI to below 25, he was exercising regularly, had stopped smoking, and his total cholesterol was below 200, but his triglyceride levels remained well above 700 mg/dL. At this point, the physician inquired further, and the patient noted that a lipid layer was regularly detected on his blood draws during his military medical exams as a young naval officer. When asked if other family members had been told similar things about their blood samples, the patient responded that he did not know. The physician tentatively diagnosed his condition as familial hyperlipoproteinemia, type 4, which responded exceptionally well to initiating omega-3 fatty acids. Impressed by the effect of the treatment and his physician's advice that the condition was likely hereditary, the patient contacted his three children, aged 26–38, urging them to have their cholesterol and triglyceride levels checked. One of his sons (IV-2), who was physically active and a nonsmoker, was found to have markedly elevated triglyceride levels at almost 1000 mg/dL.

Hypertriglyceridemia is fairly common in the American population. At least 5–10% of patients exhibit hypertriglyceridemia and the condition, which is a common problem for diabetics, raises the risk of atherosclerotic heart disease.[9] Familial hypertriglyceridemia has been described as polygenic and no genetic testing had been done on any member of this family. Its confirmed occurrence in two generations, and the presence of additional relatives with suggestive symptoms (not described in the case) implies that a single genetic variant may be associated with hypertriglyceridemia in this family. For example, an autosomal dominant inheritance pattern would be consistent with the two adult children showing no sign of hypertriglyceridemia if they simply did not inherit the putative predisposing variant. This discontinuous pattern of symptoms (in this case, significantly elevated triglycerides vs. no triglyceride elevation among siblings) is common with monogenic phenotypes. The observation also illustrates the value of obtaining information about apparently asymptomatic relatives. Most importantly, the dramatic response to omega-3 fish oils revealed a treatment that is likely to be successful in family members who develop hypertriglyceridemia in the future, regardless of whether or when a genetic cause is found.

If in fact, the familial hypertriglyceridemia arose from a known family-specific variant, this would allow early genetic testing in the future. A gene panel for lipid regulating variants might identify pathogenic variants, if they were present and described in the literature, but it is highly likely that the most common findings would be either no variant found or the identification of a VUS, as was noted for hereditary cancer syndromes and for the cardiomyopathy case described in Chapter 4. In this case, family health history was useful both for interpreting the diagnostic results and for indicating treatment in other family members.

A clinical evaluation of variants in genes associated with hypertrophic cardiomyopathy has shown that pathogenic variants are found in only a fraction of the reported candidate genes for hypertrophic cardiomyopathy and that many of the variants in those genes associated with hypertrophic cardiomyopathy are nonpathogenic or VUS. Further, variants in some of the candidate genes were associated only with syndromic conditions that included hypertrophic cardiomyopathy.[10] Another recent study has reported that genetic testing for hypertrophic cardiomyopathy variants may be cost-effective when a family's health history indicates a susceptibility.[11] At some future time, as more information is gathered about individual variants and clinical presentation in carriers, the gap between genetic testing and medically actionable diagnosis will lessen, but it is evident that for cardiovascular disease risk, family health history provides essential and cost-effective information *now* for managing adult patient health and reducing the probability of serious familial health events.

From the population study of gene-environment interactions and cardio-vascular disease risk described earlier in the chapter, the general benefit of a healthy lifestyle is evident but it is important to recognize that generalizations about the combined effects of genetic and environmental factors cannot be assumed when assessing disease risk. The hypertriglyceridemia case above illustrates that healthy lifestyle choices alone may not be sufficient per se to lessen a chronic disease risk. An intervention that targets a specific part of the overall metabolic process was necessary to alleviate the dysregulated triglycerides seen in the family described above.

The health effects of controllable environmental factors in individuals and families are not entirely understood

Smoking and obesity are frequently cited as risk factors for many chronic diseases and the conventional wisdom about their effects on health status proven repeatedly, but even for these obvious risk factors, their effects on disease risk vary. Smoking increases lung cancer risk substantially, and this causal relationship is confirmed by the signature of genetic mutations in biopsy specimens, which are consistent with those induced by tobacco smoke. However, the effect of smoking is not entirely predictable, and its effects on human physiology are varied. For instance, smoking in carriers of pathogenic *BRCA1* or *BRCA2* variants registers no significant change in breast cancer risk.[12] The proposed (and untested) explanation is that smoking reduces estrogen levels in women prior to menopause, which could reduce the hormone's growth stimulating effects in breast tissue. Another surprising epidemiological study reported that smoking partially offsets the risk of developing Parkinson's disease in subjects with a family health history of Parkinson's disease, and that alcohol and caffeine consumption ameliorate their risk for Parkinson's disease.[13]

The possibility that smoking does not further increase breast cancer risk in *BRCA1* or *BRCA2* carriers and may measurably reduce a familial-based risk for Parkinson's disease is *not* intended to equivocate about the health dangers of smoking tobacco. However, the example illustrates again that environmental factors can influence disease rates via a variety of mechanisms and therefore, their effects must be tested empirically for different health conditions. It further argues that understanding the mechanistic impact of common environmental risk factors will require further investigation. In fact, the reduction in Parkinson's incidence associated with smoking has led to efforts to explore the basis for its possible neuroprotective features.[13]

The world obesity epidemic has similarly led to more rigorous evaluation of the genetic and environmental contribution for another widely recognized risk factor, elevated body mass index. A longitudinal study of a Norwegian (i.e., genetically homogeneous) population further established the obvious and predictable conclusion: increased caloric intake increases weight in everyone, regardless of their genetic tendency to gain weight, though the effect of higher caloric intake on body mass index is greatest among those who are most genetically susceptible to elevated body mass index.[14] The authors concluded that a similar relationship likely exists in other populations. However, the study actually lays the groundwork for a host of related questions which need to be addressed further (initially laid out in Chapter 1) related to individual and familial sources of variability that can be obscured at the population level. For instance, if we examine several populations, would we find that there are novel genetic variants specific to each that affect body mass index? The discovery of an "obesity" variant in an East African population provides evidence of this possibility in subpopulations or families.[15] Do body mass index raising variants exert a substantial effect on health status that could be detected in a family health history? More generally, would we find specific health risks for those who are predisposed to weight gain and whose caloric intake is high? By contrast, would we find that someone who is not genetically prone to weight gain, but whose intake leads to a high body mass index, is as likely (maybe even more likely?) to suffer the health consequences? Most importantly, what interventions would be most effective for these subcategories of genetic risk and caloric intake? As the varied effects found from smoking illustrate, it is not a foregone conclusion that higher caloric intake is solely responsible for the health problems attributed to obesity among individual family members. It also follows that the most effective intervention might vary depending upon a given family's history for a specific chronic condition (e.g., stroke, type 2 diabetes, cardiovascular disease).

Using family health history to assess future health risk

Throughout this discussion, the cases and explanations continually point out that even when discrete genetic variants are associated with a chronic condition, the connection is obscured by other genetic variants and modifiable environmental and lifestyle factors. Nevertheless, the connection persists strongly enough that an estimate of patient risk is often possible even with a modest amount of information. Consider Table 1 which lists the risk that an individual will experience a myocardial infarction based solely on the age at which one or both parents had a myocardial infarction compared to a patient with no family health history of myocardial infarction.[16,17] This information alone is sufficient for the provider to recommend sensible interventions and treatments to reduce the risk of myocardial infarction. A retrospective study has further shown that a substantial proportion of sudden heart attack victims have previously experienced silent myocardial infarctions, which had not evoked detectable symptoms, but left scar tissue. To the extent that such myocardial infarctions may stem from a familial vulnerability, the ability to detect at-risk patients could improve substantially as the ability to diagnose and/or predict all types of cardiac events improves.[18]

TABLE 1 Age of parental myocardial infarction on risk in offspring. Odds ratio (OR) for combinations of parental heart attack history.

No family history 1.00	OR (95% CI)
One parent with heart attack ≥ 50 y of age	1.67 (1.55–1.81)
One parent with heart attack < 50 y of age	2.36 (1.89–2.95)
Both parents with heart attack ≥ 50 y of age	2.90 (2.30–3.66)
Both parents with heart attack, one < 50 y of age	3.26 (1.72–6.18)
Both parents with heart attack, both < 50 y of age	6.56 (1.39–30.95)

Reprinted with permission from American Heart Association Statistics Committee; Stroke Statistics Subcommittee. Heart disease and strokes-2016 update: a report from the American Heart Association. Circulation. 2015;133(4):e38–e360. Epub ahead of print. https://doi.org/10.1161/CIR.0000000000000350. © 2015 American Heart Association, Inc.

Physiological measures sometimes reveal trends that precede disease onset and therefore, can signal the onset of a deteriorating condition. Type 2 diabetes offers a straightforward example of this connection because diagnosis is based on a clinical test result—blood glucose levels—a physiological measurement which, in turn reflects both genetic and environmental factors. There likely are a number of genetic factors that increase the risk for type 2 diabetes, and it is well-established that if an individual has a single first degree relative with type 2 diabetes, they are three times more likely to develop it themselves. In addition, the odds dramatically increase as the number of affected relatives increase. From a management standpoint, regular monitoring of blood glucose levels among asymptomatic members of a family with a history of type 2 diabetes provides an opportunity to detect prediabetes, a potentially reversible health state that has a significant likelihood of going on to develop type 2 diabetes.[19] Similarly, a positive family health history of hypertension and hypercholesterolemia are strong predictors, not only for finding abnormal levels in other relatives, but for the secondary disease risks that follow from them,[19] but these were not predictive by themselves for type 2 diabetes risk.

The relationship between family health history and clinical test results which precede a change in health status is altogether logical when one thinks about clinical signs that precede an actual health event. Viewed in this way, the direct effects of a genetic variant are likely to appear as unusual or abnormal test results that signal the risk of a subsequent negative outcome. The observation is a recapitulation of a repeatedly mentioned principle: Gene variants almost never "cause" disease onset. Rather, variants make a negative health outcome more likely by altering the dynamics of ongoing, normal cellular and bodily processes, notably cellular growth and nutrient regulation.

Hypertension is a generalized precursor and risk factor for a variety of adverse health events. Aggressive drug treatment to reduce mean systolic blood pressure to less than 130 mm in some patients with moderate hypertension reduces the rate of cardiovascular disease and stroke substantially.[20] Nevertheless, blood pressure is often volatile and the underlying physiological conditions that bring about the chronic elevation of blood pressure with age are still unclear. Do the underlying causes and triggers of hypertension vary among unrelated individuals and by contrast, are the triggers similar among family relatives? Are the adverse health events resulting from hypertension similar for family members? The answers to these questions are currently unclear. Future studies may show whether these distinctive profiles are the result of genetic differences and subsequent cellular and system-level disruptions and whether they have a discernible effect on certain disease risks. The full utility of blood pressure readings for diagnosis and treatment may depend upon compiling a more detailed profile concerning the risks of later disease when moderate hypertension develops at a relatively early age.[21]

The scenario which has evolved about hypertension may exemplify the need to understand the biology that underlies clinical presentations. A considerable investment in the development of new technologies, tests, and diagnostic biomarkers has been made in recent years. While these tests certainly will be increasingly important for evaluating patients showing indications of chronic disease, they will also become increasingly important for evaluating the status of healthy and asymptomatic patients.

Of course, the starting point for the provider is the patient, who must then sort out the inherent and environmental components for a variety of measurements to arrive at an individual risk assessment. Several patient measures have been identified and refined over the years which are indicative of subsequent risk for cardiovascular disease. Heritability is a measure of the genetic contribution to measurements like these. As the level of heritability increases, so does the total effect of genetic factors on the measurement (i.e., the phenotype, Table 2; ref.[16] and references therein). It may seem counterintuitive that the majority of the variability in cholesterol levels is attributable to heredity given the well-known association between diet and cholesterol levels. Perhaps inherited cholesterol trends are most likely to be ascertained among family members sharing dietary and activity tendencies, but such tendencies are less apparent when looking at patients individually. As Table 2 shows, some measures such as C reactive protein levels are modestly less heritable, and these may prove to be particularly useful for assessing risk associated with dietary and other environmental risk factors. Nevertheless, if a family specific variant (even one that cannot be pinpointed) alters any specific one of these measures, then the heritability for that phenotype will be much higher among family carriers than those calculated for populations. Such relationships have not been delineated clearly enough to think of these measurements as primarily genetic or environmental, and both factors are at play in all of them. Taking the logic one step further, however, it seems likely that a pattern of metabolic measurements is most likely to prevail among family members—who share both genetic and environmental tendencies. By corollary, specific measures may be especially relevant for assessing members of a family that has shown a disease vulnerability.

TABLE 2 Heritability of clinical risk factors associated with cardiovascular disease.[a]

Cardiovascular disease risk factor	Heritability
Ankle-brachial index	0.21
Systolic blood pressure	0.42
Left ventricular mass	0.24–0.32
Body mass index	0.37–0.52
Diastolic blood pressure	0.39
Waist circumference	0.41
Visceral abdominal fat	0.36
Subcutaneous abdominal fat	0.57
Fasting glucose	0.34
C-reactive protein	0.30
Glycosylated hemoglobin	0.27
Triglycerides	0.48
High density lipoprotein	0.52
Total cholesterol	0.57
Low density lipoprotein	0.57
Glomerular filtration rate	0.33

[a]As reported in ref.[16] and references therein.

The matter of familial patterns for these types of clinical measures is not resolved at this time, including the possibility that a family may display a high level of heritability for a specific subset of these measurements, as illustrated by the aforementioned hypertriglyceridemia case.

Family health history is applied genomics

Examples presented in the chapters so far have described several cases which apparently are rooted in a heritable susceptibility. As these cases illustrate, family health history often provides the greatest clarity about a course of action for a familial disease risk, especially when genetic test results are negative or uncertain. When a familial disease pattern leaves some members experiencing the full (and often severe) impact of a chronic disease while others are asymptomatic, there is reason to consider the possibility that a hereditary condition is involved. However, because what we don't know about the genome far exceeds what we do know at this time, seeking a genetic "cause" may be expensive and futile. As noted earlier, we don't know what the effect is for many of the known variants that lie within the 2% of the genome that is intragenic, and we also know little about the function of almost 90% of human genes. Dr. Muin Khoury, the Director of the Office of Public Health Genomics at the Centers for Disease Control and Prevention raised concerns about genetic testing several years ago with physicians, encouraging them to communicate with their patients about the limitations of genetic testing[22]: "Family health history is a very informative and inexpensive 'genomic test' that can be used right now. It reflects genes, behaviors, lifestyles, and environmental factors that are shared among relatives ... Family health history can help healthcare providers assess the presence of many genetic conditions and whether patients and their relatives may have an increased risk for specific diseases." Several of the cases described in these chapters or reported in various news media reports have highlighted the need for informed deliberation about ordering genetic tests, analyzing genetic test results, interpreting what they mean, and acting upon them.[23,24] In all cases, a family's health information provides a valid basis for reaching decisions surrounding genetic tests. The challenge in medical practice now centers on developing a clinical flow that obtains family health history information, analyzes it, and then offers actionable and evidence-based recommendations to the patient, including the appropriate use of genetic tests, which are then acted on by the patient and other affected family members.

Summary and conclusions

Chronic disease risk reflects the summation and interplay of genetic variation and environmental factors. Most environmental risk factors are avoidable or ameliorated by appropriate intervention. Family health history provides a convenient and cost-effective strategy for identifying at-risk patients and offering strategies to reduce the risk and to pursue genetic testing if appropriate. However, even when genetic testing has the potential to provide clarity, the result is unclear. The utility of polygenic risk scores may improve the capability for assessing individual patient risk, though the approach does not account for the possible impact of family-specific genetic variants. Because a change in health status is the culmination of an ongoing degenerative process, and because a broad range of clinical measurements may actually be more directly tied to genetic variation, common measurements such as blood cholesterol and blood pressure may have value for distinguishing family members showing early indications of a familial disease risk from other family members who do not, even when a genetic variant cannot be identified.

- Disease risk is a product of genetic variation and environmental risk factors that usually vary for specific families, and can be discerned through a family's health history.
- Some widely used clinical measures provide an indication of underlying genetic risk and the potential for eventual disease onset that is most readily accessed through family health history.

- Genetic tests associated with clinical indications of disease risk may have utility for disease risk assessment in certain cases, but family health history is useful especially when genetic test results are unclear or negative.
- One's family members often share similar physiological and clinical tendencies that are indicative of a later disease risk, even if a specific genetic factor cannot be ascertained.
- A dichotomous distinction in quantitative clinical measures between family members may indicate the effect of an inherited genetic variant even when it is not feasible to search for it.

All three authors contributed to the content of this chapter.

References

1. Zöller B, Ohlsson H, Sundquist J, Sundquist K. Family history of venous thromboembolism is a risk factor for venous thromboembolism in combined oral contraceptive users: a nationwide case-control study. *Thrombosis J*. 2015;13:34. https://doi.org/10.1186/s12959-015-0065-x.
2. Said MA, Verweij N, van der Harst P. Associations of combined genetic and lifestyle risks with incident cardiovascular disease and diabetes in the UK Biobank Study. *JAMA Cardiol*. 2018;3:693–702. https://doi.org/10.1001/jamacardio.2018.1717.
3. Desikan RS, Fan CC, Wang Y, et al. Genetic assessment of age-associated Alzheimer disease risk: development and validation of a polygenic hazard score. *PLoS Med*. 2017;14(3):e1002258. https://doi.org/10.1371/journal.pmed.1002258.
4. Ray T. https://www.genomeweb.com/business-news/new-consumer-genomics-offerings-alzheimers-risk-come-market-despite-provider; 2018 (Accessed 31 August 2018).
5. Jansen IE, Savage JE, Watanabe K, et al. Genome-wide meta-analysis identifies new loci and functional pathways influencing Alzheimer's disease risk. *Nat Genet*. 2019;51:404–413. https://doi.org/10.1038/s41588-018-0311-9.
6. Kunkle BW, Grenier-Boley B, Sims R, et al. Genetic meta-analysis of diagnosed Alzheimer's disease identifies new risk loci and implicates Aβ, tau, immunity and lipid processing. *Nat Genet*. 2019;51:414–430. https://doi.org/10.1038/s41588-019-0358-2.
7. Cannon-Albright LA, Foster NL, Schliep K, et al. Relative risk for Alzheimer disease based on complete family history. *Neurology*. 2019;92(15):e1745–e1753. https://doi.org/10.1212/WNL.000000000000723.
8. Last AR, Ference JD, Menzel ER. Hyperlipidemia: drugs for cardiovascular risk reduction in adults. *Am Fam Physician*. 2017;95:78–87.
9. Yuan G, Al-Shali KZ, Hegele RA. Hypertriglyceridemia: its etiology, effects and treatment. *CMAJ*. 2007;176:1113–1120.
10. Ingles J. Evaluating the clinical validity of hypertrophic cardiomyopathy. *Genes Circ Genom Precis Med*. 2019;12:e002460. https://doi.org/10.1161/CIRCGEN.119.002460.
11. Catchpool M, Ramchand J, Martyn M, et al. A cost-effectiveness model of genetic testing and periodical clinical screening for the evaluation of families with dilated cardiomyopathy. *Genet Med*. 2019. https://doi.org/10.1038/s41436-019-0582-2.
12. Ginsburg O, Ghadirian P, Lubinski J, et al. Hereditary Breast Cancer Clinical Study Group. Smoking and the risk of breast cancer in BRCA1 and BRCA2 carriers: an update. *Breast Cancer Res Treat*. 2009.114:127–135. https://doi.org/10.1007/s10549-008-9977-5.
13. Ascherio A, Schwarzschild MA. The epidemiology of Parkinson's disease: risk factors and prevention. *Lancet Neurol*. 2016;15:1257–1272. https://doi.org/10.1016/S1474-4422(16)30230-7.

14. Brandkvist M, Bjørngaard JH, Ødegård RA, Åsvold BO, Sund ER, Vie GA. Quantifying the impact of genes on body mass index during the obesity epidemic: longitudinal findings from the HUNT study. *BMJ*. 2019;366:l4067. https://doi.org/10.1136/bmj.l4067.

15. Chen G, Doumatey AP, Zhou J, et al. Genome-wide analysis identifies an African-specific variant in SEMA4D associated with body mass index. *Obesity (Silver Spring)*. 2017;25:794–800. https://doi.org/10.1002/oby.21804.

16. American Heart Association Statistics Committee; Stroke Statistics Subcommittee. Heart disease and strokes-2016 update: a report from the American Heart Association. *Circulation*. 2015;133(4):e38–e360. Epub ahead of print. https://doi.org/10.1161/CIR.0000000000000350.

17. Chow CK, Islam S, Bautista L, et al. Parental history and myocardial infarction risk across the world: the INTERHEART Study. *Am Coll Cardiol*. 2011;57:619–627. https://doi.org/10.1016/j.jacc.2010.07.054.

18. Vähätalo JH, Huikuri HV, Holmström LTA, et al. Association of silent myocardial infarction and sudden cardiac death. *JAMA Cardiol*. 2019 [Epub ahead of print]. https://doi.org/10.1001/jamacardio.2019.2210.

19. Wandeler G, Paccaud F, Vollenweider P, Waeber G, Mooser V, Bochud M. Strength of family history in predicting levels of blood pressure, plasma glucose and cholesterol. *Public Health Genomics*. 2010;13:143–154. https://doi.org/10.1159/000233228.

20. Wang S, Khera R, Das SR, et al. Usefulness of a simple algorithm to identify hypertensive patients who benefit from intensive blood pressure lowering. *Am J Cardiol* 2018;122:248 254. https://doi.org/10.1016/j.amjcard.2018.03.361.

21. Yano Y, Reis JP, Colangelo LA, et al. Association of blood pressure classification in young adults using the 2017 American College of Cardiology/American Heart Association Blood Pressure Guideline with cardiovascular events later in life. *JAMA*. 2018;320:1774–1782. https://doi.org/10.1001/jama.2018.13551.

22. Ray T. *Advises Docs to Inform Patients of Limitations of DTC Gene Scans*. https://www.genomeweb.com/diagnostics/cdc-advises-docs-inform-patients-limitations-dtc-gene-scans#.XTdqRehKh3gCDC; 2010 (Accessed 11 August 2010).

23. Ray T. *Oregon Lawsuit Highlights Importance of Genetic Counseling During Period of Increasing Test*. https://www.genomeweb.com/cancer/oregon-lawsuit-highlights-importance-genetic-counseling-during-period-increasing-test-access#.XTdrEOhKh3g; 2017 (Accessed 27 October 2017).

24. Marcus AD. DNA testing was meant to help Esmé. It created turmoil. *Wall Street J*. 2019. May 17.

Chapter 7

Technology for family health history and collection and analysis

Lori A. Orlando, Brian H. Shirts, and Vincent C. Henrich

- Provider barriers to collecting and analyzing high quality family health history.
- Patient barriers to collecting and analyzing high quality family health history.
- System barriers to collecting and analyzing high quality family health history.
- Technology to overcome existing barriers to collection and analysis
- Impact on clinical care.

Chapter 2 briefly described the challenges with the current state of family health history in clinical practice. In this chapter we will review in more detail some of the existing barriers across the three stakeholders: providers, patients, and health system, and how technology may be useful in overcoming these.

Provider barriers to collecting and analyzing high quality family health history

When in medical school I learned about the need to collect family health history as part of the patient interview. I was told about how the medical interview should proceed and that one section of the interview included gathering family health history information. What I did not learn was how difficult collecting and using this information could be. It seemed so simple—ask your patient questions, record their answers, and develop an assessment and plan to provide the best possible care going forward. However, the simplicity of the steps belies the complexity of the task: collect data, analyze data, synthesize an action plan. Each of these is slow and time-consuming, and time was exactly what we were losing at an ever-accelerating rate.

Even during my days of training, the practice of medicine had started to change. There was less time with patients during clinical encounters, more pressure to see larger numbers of patients. It started around the time of the

Managing Health in the Genomic Era. https://doi.org/10.1016/B978-0-12-816015-2.00007-9

abrupt and somewhat catastrophic end of the HMO explosion. The rapidity of the HMO collapse led to increasing pressure, in a fee for service environment, to spend less time with each patient. Though at that time, we were still spending considerably more time with patients than is allowed now. So how did this impact the formal interview and family health history collection?

First, as previously alluded to, the perception of the value of family health history compared to other types of data collected in an interview seemed minimal. Hereditary syndromes were considered rare (and only a handful were known) and the familial level of risk was not well quantified at the time. If time was becoming more limited, something would have to be sacrificed, and family health history seemed an obvious choice. Instead of collecting a full history, we would focus on those things that we knew were really critical—early breast cancer, early colon cancer, and early heart disease. This abbreviated family health history seemed to suffice and we did not notice any untoward affects, as we focused more and more on the acute concerns of the patient in front of us and less on risk assessment. An inevitable consequence of devaluing family history collection was that our skill declined. Without repeatedly teasing out the information and thinking about it on a regular basis, we either lost the skill to quickly and precisely extract the necessary information and analyze it, or we never developed it in the way we developed our skills in diagnosing appendicitis or heart failure. Among those skills were: knowing what type of data to collect to meaningfully perform a risk assessment and how to educate patients on gathering the necessary information.

As a quick reminder the seven essential components of a high-quality family history include:

- Three generations (at a minimum parents, grandparents, and siblings)
- For each relative:
  - Lineage (maternal or paternal side of the family)
  - Gender
  - All medical conditions
  - Age of onset of medical conditions
  - Current age
  - If deceased, age of death and cause of death

The second impact of decreased time with patients was it limited the ability to analyze the family history information and synthesize an action plan to address areas of increased risk. The Framingham cardiovascular disease risk score, originally published in 1951, was the first statistical model designed to estimate disease risk and guide risk management.[1] By the 1990s the National Cholesterol Education Program was publishing guidelines based on counting risk factors. These were simple to implement in clinical practice and easy to perform at the point of care, while the patient was in the room with the provider. However, since then, there has been an explosion of validated risk models developed for diseases from cancer to diabetes. In addition, the models have gotten more complex, often presented as regression equations that can't be "counted" in the office. These risk models require information to be entered electronically to get the risk assessment results. The upside is that the models are more sophisticated, and, we believe, more accurate. The downside is we have to gather information from the medical record and/or the patient, access the risk calculator, enter the information into an electronic device (usually a computer), and then record the results back in the medical record. To do this for all diseases, or even a subset of diseases, for which risk assessment is possible, is not feasible or even practical. Especially, when the calculators are not centrally located or accessible—you have to find or download each one and use each separately. Even now with our more sophisticated electronic medical record systems, risk assessment is not supported. Currently, the only risk calculator available within the Epic medical record is 10-year atherosclerotic cardiovascular disease risk calculator.

Together these factors create a substantial barrier between providers and an accurate scalable process for risk assessment in clinical care. This is not to say that all providers inadequately collect or analyze family history or that there are not phenomenal family history clinicians out there, just that these challenges have broad impact on the practical application of family health history in clinical practice.

Patient barriers to collecting and analyzing high quality family health history

On the patient side the barriers can be substantial and fall into two categories: (1) knowing the medical histories of their relatives, particularly details such as the age of onset for a given condition; and (2) an accurate understanding of which medical condition affects a relative. While the two seem similar, they are quite different. The first is about an awareness of the medical conditions in the family. The second is about accurately reporting the type of condition.

The first has strong cultural influences. In the early 1900s, individuals rarely talked about their medical conditions with others, including family members. Reasons for keeping medical information private included: fear of being labeled, loss of status, or being a burden; assuming incorrectly that their personal health had no relevance for their family members. Over time our society has adopted a more open stance towards information sharing, perhaps too much so given the plethora of private information shared publicly on social media and other platforms! Culturally, this openness has been unevenly adopted. In many communities it is still taboo to discuss one's health, or if not taboo, discouraged. For example, many Asian cultures still limit sharing of medical information. In addition, some families are simply not close and do not care to communicate or share information with each other.

In these cases, the best that can be done is to communicate the "why"—that gathering the information is important—and highlight the value it provides for the entire family not just to the person gathering the information. In fact, multiple studies have shown that older members of the family will often undergo a full family history-based risk assessment not for themselves, but to pass on risk information to their children and other relatives. In one of the author's studies, a 93-year-old woman enrolled in family health history-based risk assessment study. Her physician wryly commented that she had better not get a genetic counseling recommendation! Clearly, she was beyond the age at which such an endeavor would benefit her, but as she clearly and concisely explained to our study team "she was only doing it for her grandkids." The permanent record of a family's medical history is a valuable resource for the entire family and should be documented and updated in permanent medium to pass down for generations. This type of document can become a living legacy for the family and can help avoid the problem of missing information for future generations.

Missing information is another barrier to gathering complete medical histories. World Wars I and II contribute significantly to missing information today. Families were often separated during the wars and many were never reconciled. In addition, a large number of young adults were killed in combat or as a consequence of civilian attacks. Dying young of violent causes, prevents them from living long enough to see what kinds of medical conditions they would have developed during a normal lifespan. Lastly, many records in Europe were destroyed during those wars. Another interesting example of missing information is when providers or a caregiver choose to withhold diagnoses from a patient. This is not common today, but was a prevalent form of paternalism in the past. If there was nothing that could be done then why worry the patient? Another contributing factor is the rapidity with which the field of medicine has evolved in recent years. Prior to the 1960s we often did not know what afflicted a patient, what might have caused their death, and did not have the means to make a diagnosis. For example, I have had many patients say the equivalent of: grandpa had some type of cancer, but we are not sure what. They would report that during an exploratory surgery grandpa was found to have cancer "everywhere" and the surgeons ceased their exploration, closed him up, and sent him home, since there was nothing that could be done. In those days and times pursuing a diagnosis in individuals with terminal illness wasn't necessary or desirable. Today we have so many imaging and tissue collection modalities that is rare to not have a diagnosis—even for those deemed terminal.

Apart from limited knowledge about a family's medical conditions, understanding the medical condition, is another barrier. The most pervasive reason for misunderstanding a medical condition, is that medical terminology is confusing, complicated, and filled with jargon not used in normal communication. To help make sense of medical data most people acquire a partial knowledge of the most common medical conditions and communicate information through this limited vocabulary. For example, the heart is an organ that frequently causes problems as we get older. The most frequent source of heart injury is atherosclerotic disease leading to a heart attack. But there are many other types of injury that can occur, the pericardial sack the heart sits in can become inflamed (pericarditis), there can be bleeding into the pericardial sack preventing the heart from pumping properly (pericardial tamponade), the heart valves can become dysfunctional preventing adequate blood flow out of the heart (stenosis or regurgitation), the heart muscle can be too thick (hypertrophic cardiomyopathy) or too thin (dilated cardiomyopathy), the electrical pathways in the heart can fail to conduct properly (arrhythmias), and so on. The mechanisms of injury vary widely from hereditary conditions to infections to ischemia. But often in the mind of a patient these are all just heart disease. It is critical to distinguish between these and often a skilled provider can determine which might have affected the relative. The types of medications, the symptoms, the length of time they were affected, and so on can all be important clues to the true medical diagnosis.

On top of the complexity of medical terminology and disease processes, there are two areas that frequently cause confusion, even among well-educated professionals. The first is the female organ system. This was discussed early in Chapter 2. Many individuals, both men and women, though women less frequently, interchange diseases of the cervix, uterus, ovary, and on occasion breast. It is not surprising, none of the organs except the breast can be seen, and are rarely thought about separately. In fact, many women believe that pap smears, which collect cells on the cervix to look for cervical cancer, screen for ovarian cancer. The second is differentiating primary cancer from metastasis. Just recently, a PhD asked me to help her get a second opinion on her mother's pancreatic cancer, a devastating disease with very poor survival. When I contacted the mother and gathered additional information about her history, it became clear that she had lung cancer that had spread to her pancreas. An entirely different medical condition, with different treatments, though unfortunately with an equally poor survival rate. Another common refrain is that dad had bone cancer. Bone cancer, as a primary tumor, is incredibly rare. Cancers in the bone are almost always metastasis from other more common cancers, such as breast or prostate. Traditionally, tumor classification has centered on the organ from which the cells originally arose. Treatment pathways began with the organ system and then diverged based on sub-classifications, such as cell type and whether tumor cells have spread to other organs. In this pathway, adenocarcinoma in the lung is treated differently than adenocarcinoma in the colon— even if they were the exact same cell type. Increasingly, medicine is becoming less focused on the site of origin and more focused on the cellular make up— specifically DNA and a few critical proteins. In this paradigm, tumor cells with an EGFR mutation are all treated with the same medication, regardless of where they came from. This makes intuitive sense, but will make it even harder to for the non-clinician to accurately report the cancer in their family history. For now, it is important to encourage patients to double check diseases when they include female organs or the heart, and if a cancer, that it is described by its site of origin and not by the other organs to which it has metastasized.

In the future, there may be easier and more accurate methods for gathering a family's medical information that will obviate the need for data "gathering." These will be discussed in the chapters at the end of this book. In the near term, one helpful approach to facilitating accurate data collection by patients is to provide them with guidance on the value of family health history collection and documentation, the tools to engage their relatives, and the time to talk with relatives and follow up on questions before having to report the results of their effort to their provider. Many families contain a family historian, the one relative that knows almost everything about everybody. Approaching this relative first can often reduce the effort needed to complete the data collection; and create a quick entry point to get started.

System barriers to collecting and analyzing high quality family health history

It may seem odd to include health systems as a stakeholder in the family health history process, but since the health system provides the supporting infrastructure for medical care, it also has the mandate to provide the supporting infrastructure for collecting and analyzing family health history data. As most of us have experienced, clinical care now occurs primarily in health settings that are owned or controlled by either the medical provider or larger health related entities. The patients come to the medical provider, not the other way around. The rationale for this shift in where healthcare is provided, is that consolidating medical services and resources in a single location improves both quality and efficiency, and thus, theoretically, health-related outcomes.

In many cases, particularly intensive care units, emergency rooms, and operating rooms co-locating resources does indeed improve health outcomes. In the office setting, where family health history-based risk assessments should be taking place, resources are inadequate to support data collection, recording, or analysis. As mentioned earlier, in the provider barriers, time has become a limited commodity, in large part, due to the demands of the health system. In addition to the time constraints imposed by the health system, the tools provided for family health history and risk assessment, particularly recording, annotating, and analyzing family medical histories, are not sufficient. Until the Hitech Act, most providers documented on paper and thus were bound to paper charts. With the Hitech's incentives and regulatory pressures, the majority of large practices and some small groups have transitioned to electronic medical records. The investment in medical record system in recent years has been astronomical with transitions from paper to electronic charts costing upwards of $1–2 billion dollars for large health systems. With this massive investment in a digital infrastructure, there was great potential for enhancing the process of data gathering and analysis, yet it remains essentially unchanged from when it was performed in a paper environment. Most electronic medical records provide a location for family health history data and some type of database; but the relatives and relative relationships are limited, the choice of medical conditions are limited, and there are no options for indicating age and cause of death. These restrictions force providers to enter much of the data as free text, which hampers attempts at automating analysis. Thus, in the current state, regardless of whether the health system has a paper chart or electronic medical record, family health history data and risk assessment continue to be actions of recording data as text, and analyzing by hand, if the provider is well informed and has the time to do so.

Beyond, simply providing the opportunity and the infrastructure to support risk assessment, health systems could support it as an important component of their patients' care. Unfortunately, in a fee for service environment, risk assessment is disincentivized. There are no tangible or intangible benefits to maintaining population health, rather than providing services when they become sick. In fact, the entire medical system is driven by managing ill patients and devotes a considerable amount of dollars and attention to improving the care of those who are not healthy. This has led to the healthcare system being dubbed the "sick care" system in some circles. If all reimbursements are tied to the interventions you perform on your patients, it is very hard to sell the idea of risk assessment and risk management to maintain health. This is not to say that providers do not want their patients to be healthy or that the healthcare system is overtly trying to encourage illness; but that in the current era policies, regulations, and reimbursement strategies have all reinforced a vicious negative cycle that focuses on the needs of those who are unwell, at the expense of those who are healthy. There is some hope that the recent shift toward value-based care and reimbursement for maintaining health may drive greater attention and investment in tools to support wellness, such as those that could streamline the risk assessment process.

Technology to overcome existing barriers to collection and analysis

Family health history collection, risk assessment, and synthesis of an actionable risk management plan can be performed efficiently and effectively using a variety of digital tools that have the potential to overcome the barriers created by a reliance on providers to gather, record, and analyze family health history data. Given the patient and provider barriers previously cited, one solution is that patients, instead of providers, should be the locus for data collection and recording. Several important features of this shift could significantly enhance the risk assessment process. First, this offloads data collection from the provider and allows them to instead focus on risk management strategies. Second, it frees the patient from the constraints of the clinical visit. Instead of needing to recite their family's medical history in detail and on demand, during their appointment, they can enter their information online when they have the time to thoughtfully collect and enter their information. Importantly, de-linking family health history data collection from the clinical encounter creates an opportunity to implement educational resources on the value of family health history for their own health and the health of their offspring, how to ask relatives about their medical history, and what types of information to gather. As previously, noted these can significantly enhance the quantity and quality of the data collected and will optimize the results of the risk assessment process. Several studies have shown high concordance between digital family health history data entered by the patient and family health data gathered as part of a three-generation pedigree by genetic counselors.[2] In addition, when compared to family health data gathered in the course of routine clinical care, digital tools are markedly superior. In one example of 1124 primary care patients, most (66%) did not have enough family history data documented in their medical record to perform a risk assessment for six common diseases, while all were able to enter all data necessary to run the risk assessment using the Family Healthware program. In addition, 23% had at least a moderately strong risk for one of the six diseases, but were not noted to be at increased risk in the medical record.[3]

Another crucial benefit of a digital risk assessment is that information entered by the patient can be stored in discrete data fields, and used to automate risk assessment and generate clinical decision support. Risk assessment algorithms are published by guideline bodies in paper format, but developers can code these guidelines and integrate them into digital tools, permitting the risk assessment process to seamlessly trigger a list of recommended risk management strategies for every individual. The ease of instantaneously coordinating this process across multiple conditions in a single location is transformative.

The ideal family health history platform

In 2014, deHoog outlined the ideal features for a digital family health history-based risk assessment: computerized, patient administered, easy to use, collects all data necessary for risk stratification, updateable, has integrated risk algorithms and evidence-based clinical decision support, and can communicate with the electronic medical record.[4] In comparison most electronic medical records are not patient administered, not easy to use, do not collect all of the necessary data, and do not have integrated risk algorithms or clinical decision support. Refer back to the section on health system barriers for more details on the limitations of current electronic medical records.

At the time of writing several digital solutions have been developed though, as of yet, none have all the "ideal" characteristics. A number were developed to complement services offered by a company or were purchased by companies to assist them in driving demand for their product. For example, CancerGene Connect was acquired by Invitae and Myriad built their own. This leaves a significant gap in the space where family health history-based risk assessment is most successfully addressed—primary care and cancer clinics. Table 1 is a list of several digital tools available today.

TABLE 1 A listing of some electronic family health history collection programs.

FFH tool	Number of diseases	Decision support provided to	Supports HL7 inter-operability	Availability to patients	Affiliated with company
CancerGene Connect	112	Clinician	✓	Through HCP	Invitae
CancerIQ Self-Assessment	30	Clinician	✓	Through HCP	
CRA Health	18	Clinician	✓	Through HCP	✓
Family Healthware	6	Patient		Available to public for $9.99/mo	Sanitas
Health Heritage	47	Patient	✓	Through HCP	NantHealth
Inherited Health	282	Patient		Available to the public for $39.95/yr	Informed DNA
Invitae FHx Tool	24	Clinician	✓	Through HCP	✓
ItRunsInMyFamily	97	Patient	✓	Available to public for free	
MeTree	123	Clinician and patient	✓	Research study access only	
My Family Health Portrait	87	Patient	✓	Available to public for free	
MyFamilyHealth	15	Patient		Available for free. Not fully functional.	
MyLegacy	12	Clinician and patient	✓	Through HCP	✓
Myriad FHx tool	26	Patient		Available to public for free	✓
Progeny FHQ	387	Clinician	✓	Through HCP	✓
VICKY	20	None	✓	Research study access only	

Reprinted with permission from Elsevier. Ginsburg GS, Wu RR, Orlando LA. Family health history: underutilized for actionable risk assessment. *Lancet.* Published online August 5, 2019. https://doi.org/10.1016/S0140-6736(19)31275-9. PMID: 31395442.

The single biggest challenge for digital family health history tools has been connecting to electronic medical records. Fortunately, new government regulations for meaningful use and interoperability of electronic medical records, and emerging informatics data standards, are beginning to solve this problem. Updates to regulations supporting the premise that patients own their data, are forcing electronic medical records to open their programs and allow patients access to their data. This access permits them to retrieve and share it with their own personal health record or third-party applications, such as a digital risk assessment tool. In informatics, health-related entities have adopted standards for seamless integration, authentication, and authorization from other industries and incorporated them with standard health-related data terminologies to yield SMART on FHIR (Substitutable Medical Applications, Reusable Technology on Fast Healthcare Interoperability Resource).[5]

An application that is SMART compatible can connect with another application (in this case an electronic medical record) using the same mechanisms that allows you to use your Google login to access your Pinterest account. Essentially, you are (1) telling the medical record that you have an account with the other application, (2) that you are who you say you are, (3) that it is okay to share information, and (4) that you can keep this connection open and have "views" from one application inside the other. An example of the "views" is when you can watch a YouTube video that is embedded in a Google search page, without having to go to the YouTube site. In effect you can view pages from the other application, such as a risk assessment tool, within the medical record—allowing the other application to interact with the provider in their normal workflow within the medical record. This connection can be created for the patient—so they can access and send data from their medical record to another application; and it can be created for the provider—so they can see the results of a digital risk assessment within the medical record, minimizing workflow interruptions.

An application that is FHIR compatible allows two applications to view, manipulate, and interpret data in the same way. Essentially it constrains the meaning of a data element so that all parties agree that it means the same thing and allows it to be shared between applications via an application programming interface (api). For example, if the medical record says that the patient has right invasive ductal carcinoma and that the mother had breast cancer without metastases, both should be processed by the risk calculator as the same thing—the patient herself had breast cancer and her mother had breast cancer as well. Using medical terminologies, such as SNOMED-CT (Systematized Nomenclature of Medicine-Clinical Terms), LOINC (Logical Observation Identifiers Names and Codes), and others, which associate codes with terms, allows compatibility across different applications so that no matter where the data comes from, it means the same thing.

Ultimately, SMART on FHIR allows software developers to do, what they can do much better than the electronic medical record vendors, leverage graphical user interfaces for usability, facilitate patient entry of data, and visualize data to promote knowledge assimilation and clinical decision support for providers.

Impact on clinical care

In the optimal scenario family health history data collection is removed from the actual clinic visit, moving from the traditional clinical model (Fig. 1A) to the optimal clinical model (Fig. 1B).

In the optimal model, patients, empowered and educated on how to gather high quality and thorough family history information, confer with relatives then access a patient-facing platform where they enter their family data. Providing up-front patient education about family health history and what information is

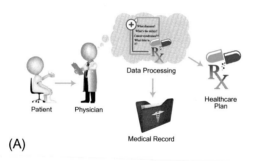

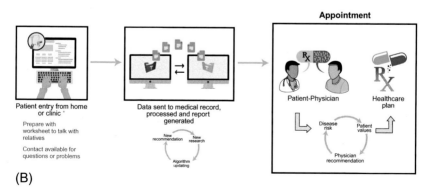

FIG. 1   (A) Traditional flow of family health history data. (B) Ideal flow of family health history data.

important to collect, is critical, and has been shown to improve patients' ability to provide complete and accurate information that impacts the conditions they are identified as at risk for. After patients enter all information, the platform automates the running of risk algorithms, and provides clinical decision support to both patient (in real time) and provider (at the point of care) in order to facilitate shared-decision making. As a proof-of-concept for this model, MeTree was integrated with the Epic electronic medical record at Duke University via a SMART on FHIR connection. This successful demonstration, funded by the National Human Genome Research Institute's Implementing Genomics In Practice (IGNITE) network, allows patients to access the risk assessment platform through a single sign-on link in the patient portal, pulls relevant data into the platform to prepopulate fields, and generates a graphical dashboard with clinical decision support recommendations for providers within the patient's chart. With this final hurdle addressed, the potential for technological solutions to meet all of the deHoog criteria, and revolutionize systematic risk assessment, is at hand.

The questions you may be asking yourself are, "How much of a difference is this really going to make?," "Will patients really do this?," and "Isn't this going to derail my conversations with patients, if I want to focus on getting their diabetes under control and they want to talk about their risk assessment?." Given how important these are to the ability to perform clinical care let's review each one separately.

Patient's ability and willingness to enter family health history information into a digital health risk assessment tool

Several studies have shown that patients, when offered the opportunity to use tools like these, will do so. Studies such as the Family Healthware Trial and our work with MeTree have shown that patients are willing and able to collect data and enter it into a digital tool. The Family Healthware Trial recruited 3786 patients from 41 primary care practices in 13 states and the MeTree studies have recruited 5500 patients from eight national healthcare systems and two international healthcare systems. In addition, surveys and interviews of patients recruited in the MeTree trials found that the platform was easy to use (93%), easy to understand (97%), useful (98%), raised awareness of disease risk (81%), and changed how they think about their health (82%).[6] In addition, 66% talked with relatives (on average 2.2 relatives) to better understand medical conditions that run in the family.[7] Mean completion time has been ~ 25 min, which does not take in to account time spent collecting information from relatives. Given the significant time commitment required to capture a complete family health history, it is clear from our own work and from others that patients, not providers, need to serve as the main locus for data input.

Equitable access to digital tools is a concern that faces all genomic interventions, but is particularly important to address in the face of technology solutions for family health history-based risk assessments. If low resource, low literacy populations are unable to benefit from these technologies then there is a potential that they will widen disparity gaps rather than close them. This has already been seen in low-literacy patients trying to navigate <u>My Family Health Portrait</u>, which prompted investigators to create an animated relational agent (VICKY) to pose verbal questions that users answer by selecting from text-box responses.[8] With VICKY, low literacy minority users entered more data, than without. However, this format is not effective for collecting large numbers of conditions and doesn't address the other types of help needed. Studies of MeTree found that completion rates varied by education: 70% (high school), 76% (community college), and 83% (4-yr college); though engagement (talking with relatives, perceived value) was twice as high in the low literacy group. Notably there were no differences by ethnicity or race. To understand better what literacy-related barriers exist we studied 20 patients (56% female, 76% Lumbee Indian, 12% African American, 40% living at poverty level, 30% with high school or less education, 88% with internet access). During the pilot, a consistent pattern emerged. All were highly motivated, gathered family health history from relatives (mean number relatives 2.7), and entered more than the required 6 relatives (mean 11), but many had difficulty navigating the data entry interface (e.g., selecting appropriate disease from drop down lists organized by organ system). Given time they could do it, but it was not intuitive for them, suggesting that text-based data entry interfaces are unlikely to work well for those with low health literacy. With this feedback and VICKY's successes, we believe that to ensure that risk assessment is broadly available and easy to use that these types of tools should support interfaces with natural language processing and machine learning to allow voice-to-text response capture from mobile devices, including phones. Essentially creating a "Siri"-like experience for family health history—say "mom" and "kidney stones" and the data is entered.

Provider's perceptions of risk assessment technology and its impact on their clinical care

There are not many studies that explore the impact of integrating a digital risk assessment in primary care. In our studies with MeTree, providers indicated that it improved their practice (86%), improved their understanding of family health history (64%), made practice easier (79%), did not disrupt their planned conversations with patients (69%), and was worthy of recommending to their peers (93%).[6] In addition, most providers prior to study start were skeptical that the digital risk assessment would change their practice. Most believed that they were already identifying high risk patients, and that there would not be many "new" ones found after implementing a digital risk assessment. As you will read below, the opposite was true. We found a considerable proportion of the population was meeting criteria for higher risk management interventions. The ultimate impact of these results was that providers requested the health system to continue support for the clinical trial beyond the funding period. While anecdotal, this experience indicates that done properly, digital risk assessment can enhance and support providers in caring for their patients, with minimal disruption to their workflow.

Does systematic risk assessment using digital health technology enhance the quality of clinical care

In studies of systematic family health history-based risk assessments in unselected populations, 24–50% of patients meet risk criteria for (actionable) hereditary conditions depending upon the number and type of conditions assessed.[3,9,10] These were the first studies to show that systematic assessment of *multiple conditions at once* was efficient and effective. Previously, most studies of risk assessment focused on single conditions, an approach that is not scalable in primary care where patients are at risk for a wide variety of conditions. In a community-based pilot study of a patient-facing web-service with integrated evidence-based clinical decision support (MeTree), 25% of 1184 participants met criteria for genetic counseling for Lynch and Hereditary Breast and Ovarian Cancer, and 19% met criteria for enhanced cancer screening (screening that differs from population-based guidelines).[11] Of these, only 6% had received recommended care prior to MeTree. After using MeTree 79% received recommended care. Similarly, in an NIH funded study "Implementation, Adoption, and Utility of Family History in Diverse Care Settings," which integrated MeTree into primary care clinics at 5 geographically and culturally diverse healthcare systems, 20% of 2514 participants met criteria for cancer genetic counseling, 6% for cardiac genetic counseling, 22% for enhanced cancer screening, and 4% for familial hypercholesterolemia testing. These data strongly support the value of family health history based risk assessments, demonstrating their utility and potential to improve population health when applied systematically to the general population.

Summary

If digital technology is to be implemented into clinical care, one question remains: how to integrate the process. We have explored several models and obviously, each setting is unique and have unique solutions; but one method for ensuring patients have access, education, and results would be to incorporate these tasks into the clerk's and clinical nurses' repertoires. When patients schedule appointments, clerks can indicate that providers would like the risk assessment completed prior to their appointment, and the clerks can send an access link or describe how to access through the patient portal. When completed nurses could be notified by the platform via messaging through email or the electronic medical record. Nurses could review results at triage when patients arrive for their appointment and if any actionable risk management strategies are recommended the nurse notifies the provider. There are of course, other potential paths for integration and in the last chapter we talk about the potential for health systems to take over the task of systematic risk assessment, but in the setting of the clinical care encounter engaging support from staff will further streamline the workflow.

Lori A. Orlando is the lead author for this chapter. First person statements are from her point of view.

References

1. Dawber TR, Meadors GF, Moore Jr. FE. Epidemiological approaches to heart disease: the Framingham study. *Am J Public Health Nations Health*. 1951;41(3):279–281.
2. Reid GT, Walter FM, Brisbane JM, Emery JD. Family history questionnaires designed for clinical use: a systematic review. *Public Health Genomics*. 2009;12(2):73–83. https://doi.org/10.1159/000160667.
3. O'Neill SM, Rubinstein WS, Wang C, et al. Familial risk for common diseases in primary care: the family healthware impact trial. *Am J Prev Med*. 2009;36(6):506–514. https://doi.org/10.1016/j.amepre.2009.03.002.
4. de Hoog CLMM, Portegijs PJM, Stoffers HEJH. Family history tools for primary care are not ready yet to be implemented. A systematic review. *Eur J Gen Pract*. 2014;20(2):125–133.
5. Mandl KD, Mandel JC, Kohane IS, Kreda DA, Ramoni RB. SMART on FHIR: a standards-based, interoperable apps platform for electronic health records. *J Am Med Inform Assoc*. 2016;23(5):899–908. https://doi.org/10.1093/jamia/ocv189.
6. Wu RW, Orlando LA, Himmel T, et al. Patient and primary care provider experience using a family health history collection, risk stratification, and clinical decision support tool: a type 2 hybrid controlled implementation-effectiveness trial. *BMC Fam Pract*. 2013;14:111. https://doi.org/10.1186/1471-2296-14-111.
7. Beadles CA, Ryanne Wu R, Himmel T, et al. Providing patient education: impact on quantity and quality of family health history collection. *Fam Cancer*. 2014;13(2):325–332. https://doi.org/10.1007/s10689-014-9701-z.
8. Wang C, Bickmore T, Bowen DJ, et al. Acceptability and feasibility of a virtual counselor (VICKY) to collect family health histories. *Genet Med*. 2015;17(10):822–830. https://doi.org/10.1038/gim.2014.198.
9. Cohn WF, Ropka ME, Pelletier SL, et al. Health heritage, a web-based tool for the collection and assessment of family health history: initial user experience and analytic validity. *Public Health Genomics*. 2010;13(7–8):477–491. https://doi.org/10.1159/000294415.
10. Orlando LA, Wu RR, Beadles C, et al. Implementing family health history risk stratification in primary care: impact of guideline criteria on populations and resource demand. *Am J Med Genet C Semin Med Genet*. 2014;166(1):24–33. https://doi.org/10.1002/ajmg.c.31388.
11. Orlando LA, Wu RR, Myers RA, et al. Clinical utility of a web-enabled risk-assessment and clinical decision support program. *Genet Med*. 2016;18:1020–1028. https://doi.org/10.1038/gim.2015.210.

Chapter 8

Family health history and genetic counseling

Rachel A. Mills, Lori A. Orlando, Brian H. Shirts, and Vincent C. Henrich

- Genetic counselors are allied health professionals with special training in genetics and counseling.
- Discussion of family health history can reveal information about family dynamics, available support systems, and other issues.
- It is important for any practitioner providing genetics services to support and respect patient's decisions and their decision-making process.
- Genetic counselors can determine which individual in a family should be tested.
- Genetic counselors' expertise in understanding different types of testing can help determine the most appropriate test to answer the diagnostic question.
- Genetic counselors play an important role in helping patients understand the possible outcomes of the test and, when results are available, what those results mean.
- Genetic counselors and policy makers are considering new ways to provide genetic counseling services that will meet the growing needs of patients and their referring providers.

So you've collected family history information from your patient. Now what? A previous chapter touched on clinical guidelines providers can follow to determine next steps in treatment or prevention. Some of those guidelines may direct you to order genetic testing for your patient. But if the thought of keeping up with just one more professional guideline or navigating the abundance of available genetic tests has you worried, consider seeking the help of a genetic counselor.

What is a genetic counselor?

Genetic counselors are allied health professionals with special training in genetics and counseling. Their advanced training enables them to assess risk of genetic disease, interpret genetic test results, and support patients in information-seeking and decision-making about their genetic health. Genetic counselors are adept at taking the complex world of genetics and simplifying it

Managing Health in the Genomic Era. https://doi.org/10.1016/B978-0-12-816015-2.00008-0

to the needs of their patients. They are also the experts at collecting and interpreting family health history.

Collection of family health history information is foundational to the practice of genetic counseling. Most appointments with a genetic counselor begin with the collection of a detailed family history. While utilization of tables and lists to record family history may work for many providers, genetic counselors craft pedigrees with standardized symbols—circles are females, squares are males. These standards are regularly revisited and updated; for example, some genetic professionals have called for revision of pedigree nomenclature to reflect transgender, intersex, and non-binary patients. You've already seen a number of examples of pedigrees in previous chapters (Fig. 1).

The genetic pedigree is an icon of genetic counseling. Amazing discoveries have been made thanks to the power of the pedigree. A form of ALS affecting groups of Appalachian families and caused by a gene called *FUS* has been extensively studied by neurologist Edward Kasarskis at the University of Kentucky thanks in part to a 250-page digital pedigree consisting of close to 7000 relatives. Geneticist Mary Claire King also used large family pedigrees to determine that a gene on chromosome 17 is responsible for hereditary breast and ovarian cancer—you've probably heard of *BRCA1*, thanks in part to actress Angelina Jolie. Prior to development of rapid sequencing technologies, extensive and large family histories have been necessary to link genetic variants to disease.

When considering family history in primary care, a 250-page pedigree isn't necessary. This size pedigree is most useful for research using linkage analyses to discover genes and their association with disease. Clinical pedigrees usually

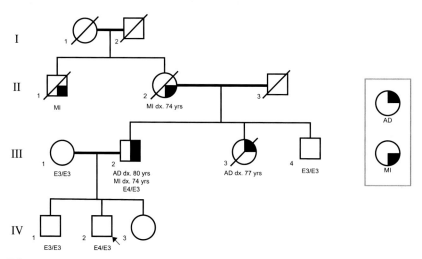

FIG. 1 Family health history for case involving APOE gene. E3 refers to *APOE*e3* variant. E4 refers to *APOE*e4* variant.

start with three generations: the proband (or patient) and his/her siblings, parents, and both sets of grandparents. Aunts, uncles and first cousins may also be included, as are the proband's children and grandchildren, if applicable. Risk of hereditary disease is greatest with this small core of family members. Unfortunately, it may not be possible to complete a 3-generation pedigree due to factors like adoption or family estrangement. Lack of information or misinformation can impact the interpretation of a pedigree and assignment of disease risk. In cases of missing information, disease risk is determined by the individual's personal history and population risk information. Pedigrees should be updated regularly (at least every 5 years, but ideally every time a new health event occurs) to note changes and new diagnoses in the family history or to correct misinformation.

Using family health history as a psychosocial instrument

Providing a family history in a medical setting may be a new experience for patients; so explaining the purpose of collecting family health history and what can be learned is important at the initiation of the conversation. Genetic counselors use collection of family history as a psychosocial instrument. **Discussion of family health history can reveal information about family dynamics, available support systems, and other issues**. Questions about family history can also bring up painful memories around loss, which can benefit from an empathetic response. Information about communication patterns and family structure that can impact a patient's care may be revealed through the process of collecting family history. For example, collecting family history information about a pediatric patient may reveal that her mother is a single parent without additional family support; this information could provide insight about why the patient often misses appointments (her mother, the sole provider for the family, cannot take off of work), and would be an important piece of information, if a genetic test that requires a trio of samples (patient and both parents) is indicated. Although a counselor may not record psychosocial information in the medical pedigree, these discussions can provide insight about the patient's experience and improve trust between patient and provider.

Autonomous decision making about genetic testing

By referencing a patient's family health history, genetic counselors can use tools like Bayesian statistical analysis (discussed in Chapter 3 in reference to the evidence needed to classify a variant), diagnostic criteria, and clinical guidelines to estimate the likelihood that an individual may have inherited a genetic variation that causes a genetic disease or increases the chances of developing a particular health condition. If there is high likelihood of a hereditary disease, genetic testing may be appropriate, and genetic counselors can guide patients through the process of testing. Patient autonomy is key in the decision whether to pursue genetic testing, and genetic counselors aim to enable autonomous decision-making. In short, counselors provide the risks and benefits of undergoing testing and facilitate the patient's decision making process. Genetic testing may not be desirable to all patients, and providers have a responsibility to respect the patient's right to know or their right *not* to know their genetic information.

I learned a lesson in the importance of autonomous decision-making during a clinical rotation in a cancer clinic. A patient who had recently been diagnosed with breast cancer was given the option to undergo genetic testing, as her personal and family history indicated a risk for hereditary breast and ovarian cancer (HBOC). Due to her recent diagnosis and other issues with her family and her health, she felt that undergoing testing could lead to negative outcomes, and she "couldn't handle any more bad news." Her treatment plan would not be impacted by knowing her genetic risk of HBOC; so, the appropriate decision for her was to decline testing at that time. We provided her with resources for additional psychosocial support and scheduled a follow-up appointment where we would revisit testing options.

Some patients may decline to undergo testing as an "ignorance is bliss" approach, if they are asymptomatic. This is often observed in individuals at risk of genetic diseases that do not have treatment options, such as Huntington's disease. As discussed in Chapter 3, Huntington's disease is an autosomal dominant progressive motor, cognitive and psychiatric disorder caused by a repetition of three nucleotides (CAG) in the *HTT* gene. A review of patients in the UK at risk for Huntington's disease revealed that over 80% have not undergone predictive testing.[1] The choice to undergo testing may be influenced by the absence of treatments, anxiety about an abnormal result, personal experiences of caring for effected relatives, or perceived stigma associated with the condition. The ethics of decision-making in genetics practice have been deliberated elsewhere, so I won't belabor the discussion here. In summary, **it is important for any practitioner providing genetics services to support and respect patient's decisions and their decision-making process.**

How genetic counselors can help providers

Genetic counselors not only have counseling expertise to facilitate decision-making about undergoing testing, but they also have the necessary genetics knowledge to determine what type of genetic test is most appropriate. Alongside geneticists and other MD clinicians specially trained in medical genetics, genetic oncology, and maternal fetal medicine, genetic counselors are able to narrow the list of differential diagnoses and identify the most appropriate test to answer the diagnostic question at hand. During the counseling process, genetic counselors identify the best person in the family to test, select the optimal test and laboratory, and compare test costs, likelihood of insurance coverage and obtain insurance prior authorization.[2]

Who should be tested?

Genetic counselors can determine which individual in a family should be tested. For symptomatic patients that is usually the patient. In risk based testing, when a patient is asymptomatic and healthy, the general rule is to test the closest symptomatic relative who is able to consent and is agreeable to testing. Let's consider an example case. Your patient is a 42-year-old female who shares that her sister was recently diagnosed with colon cancer at age 47. She reminds you that her mother died from colon cancer 10 years ago. The family history of two first degree relatives diagnosed with colon cancer, one under the age of 50 should be a red flag for the potential of a hereditary cancer syndrome, specifically Lynch syndrome (also known as hereditary non-polyposis colorectal cancer, or HNPCC). In this case, the best person to test would be your patient's sister. Testing your patient would not be as informative because the interpretation of a negative result does not provide definitive information. If your patient was not positive for a mutation associated with Lynch syndrome, is it because she is truly negative and did not inherit her family's risk of cancer? Or is it because the family's cancer is caused by a genetic variant not detected by the test? Testing the patient's sister first could provide clarity to this set of questions. Additionally, if the sister's test reveals a genetic variant likely explaining the family history of colon cancer, your patient could then be tested only for the specific familial variant (a test that is faster and cheaper) rather than an entire hereditary colon cancer panel. The practice of testing family members after identifying a familial variant is referred to as cascade testing. In some cases, an affected family member is not available or does not choose to undergo testing. *Testing an unaffected individual is possible, but the limitations of testing a healthy patient, particularly the challenges in interpreting a negative result must be considered when using results to guide clinical decision-making.*

Which test to choose?

As mentioned, **genetic counselors' expertise in understanding different types of testing can help determine the most appropriate test to answer the diagnostic question.** As of 2017, there were approximately 75,000 genetic tests on the market, and about 10 new tests enter the market every day.[3] Available tests include chromosomal analysis, single gene tests, panels of multiple genes, exome analyses, and genome analyses (these were briefly described in Chapter 3). With that number of tests available it can be overwhelming to choose the best one for the patient! *Unfortunately, the rate of incorrectly ordered genetic tests is almost twice as high in non-genetics providers compared to genetics providers.* But, including a genetic counselor in the testing process can reduce the amount of unnecessary or inappropriate genetic testing, saving hundreds of thousands of dollars (Table 1).[2]

TABLE 1 The impact of genetic counselors on testing and costs.[2]

	Institutional Lab Seattle Children's Hospital (n = 3441 genetic orders, 4 yr 9 mos)	Institutional Lab Health Partners (n = 904 cases, 12 mos)	Institutional Lab Health Partners (n = 80 oncology-related cases, 13 mcs)	Reference Lab Laboratory Corporation of America (n = 280, 14 mos)
Total cost tracked (test list price)	$5,965,000	$557,758	$166,673	
Total cost savings to institution/patient	$972,000	$263,400	$153,292	$148,000
Savings per test	$282	$275	$2472	$528
Rate of test modification	32%	13.5%	9.3%	54%
Number of consults with providers			56%	
Number of cases that included investigating patient's insurance coverage		Most orders for tests ≥ $500 submitted for prior authorization	41%	
Number of referrals to genetics professionals			10%	

yr, years; mos, months.

In fact, genetic counselors are so adept at gathering and interpreting family health history and identifying appropriate genetic testing that some laboratories and insurance companies require genetic counseling before testing can be conducted and/or reimbursed.[4]

Facilitating patient understanding

Genetic counselors play an important role in helping patients understand the possible outcomes of the test and, when results are available, what those results mean. Informing patients of the different types of possible results and outcomes prior to testing, can help set appropriate expectations and better prepare patients to understand results. Genetic testing outcomes are not always as simple as "positive" or "negative." Particularly with sequencing technologies, it is also possible to identify variants of undetermined significance (VUS) (described in Chapters 3 and 4). The concept of VUS can be very confusing to patients (and many providers as well!). So preparing patients for the possibility, helps to avoid confusion once results are available. During pre-test counseling possible test results are described. A genetic counselor may say: "There are three possible results for this test: a positive result which will provide a diagnosis or an explanation for your symptoms, a negative result which means that you do not have any of the genetic variants known to cause this disease, or a variant of undetermined significance which means that you have a genetic variant, but we don't know the meaning of the variant, usually because it has not been seen in many patients before." Preparing patients for possible results does not just apply to a VUS, but also to any unexpected or confusing outcome. For example, genetic testing may reveal non-paternity or secondary findings unrelated to the reason for testing.

As mentioned at the beginning of this chapter, genetic counselors' specialized training in genetics, as well as counseling and communication make them experts in taking the complex world of genetics and simplifying it to the needs of patients. It also helps them in their role as a bridge between non-genetics clinical providers and the world of genetics! The typical model of genetic counseling is usually separated into two patient interactions: pre-test and post-test counseling. In its simplest state, pre-test counseling involves genetic counselors providing education about genetics, facilitating shared decision-making, and, if the patient elects to have testing, providing details about the test and possible outcomes. Post-test counseling usually includes interpreting results, discussing the impact of those results on care, and next steps, which may include additional testing or referral to other specialists. In some settings, such as pediatric genetics, genetic counselors may provide continuous and coordinated care for patients and their families over many years. I've described an overly simplistic view of the processes of genetic counseling here for brevity—but if you are interested in learning more about the complexities of genetic counseling, I recommend core genetic counseling texts including A Guide to Genetic Counseling[5] or Facilitating the Genetic Counseling Process: Practice-Based Skills.[6]

How do I refer to or work with a genetic counselor?

Hopefully I've made a case for genetic counselors so far! Counselors can collect and interpret family health history, provide psychosocial counseling related to genetics, and maneuver through the muddy waters of decision-making for testing. So how do you bring in a genetic counselor to help your patient? Typically, when a provider identifies a patient they think may be at risk for a genetic disease, based on personal or family health history, they refer him/her to genetic counseling as they would to any other specialist. This process is fairly straightforward for providers who are part of medical centers, as most clinical genetic counselors work in academic medical centers, or public or private medical facilities. However, referrals to genetic counseling can be tricky due to the limited number of practicing clinical genetic counselors and the amount of time it takes to do both the pre-test and post-test counseling. On average a genetic counselor may see 4 to 5 patients in a day, depending upon their complexity. This has greatly limited access to genetic counselors even in areas where that have them. In many cases, wait times for genetic counseling appointments can be many months. Often, genetic counselors work in larger metropolitan areas, which also impacts accessibility for patients in more rural settings. At this time, there are over 4000 certified genetic counselors in the United States (http://www.ABGC. net; counselors are certified if they complete a 2-year Master's level training program and pass a certification exam). Of those, only about 60% provide direct patient care[7]—the others conduct research, work in genetic testing laboratories, and write book chapters about family health history.

In the past, the majority of clinical genetics patients were individuals and families with rare, Mendelian-inherited disorders. However, as the understanding of the genetic impact on common diseases has grown, the patient population has shifted, and we are now seeing an increasing number of other patient types, including, ostensibly, healthy patients. In part, this increase is being driven by a greater understanding of and interest in genetics by both providers and patients. As described in Chapter 3 the DTC market has taken advantage of rising consumer interests and the availability of testing has further increased consumer interest. In fact, the DTC ancestry and health test offered by 23&Me was a top 5 seller during Amazon's Black Friday sales in 2017. Combined with other tests offered by Ancestry.com, MyHeritage and FamilyTree DNA, 12 million people had undergone genetic testing in 2017[8] and by 2019 23&Me and AncestryDNA had 10 million customers each. Consumer-facing genetic and genomic testing is a great way to introduce the public to the importance and impact of genetics. However, DTC testing can present a number of challenges to providers including genetic counselors.

Challenges of direct-to-consumer genetic testing

- Countless stories have been shared by consumers who learned new and unexpected information about themselves through this testing (Dad isn't actually your dad! Your mom's fertility doctor used his own sperm! And more amazingly chilling stories…).

- Some consumers have unrealistic expectations of the capabilities of DTC testing. This concern has been highlighted following 23&Me receiving approval from the Food & Drug Administration to provide some results from *BRCA* testing (reminder: *BRCA1* and *BRCA2* are the genes associated with inherited breast and ovarian cancer, HBOC).[9] 23&Me is limited to reporting three variants that are most commonly observed in individuals with Ashkenazi Jewish ancestry. However, this can lead to confusion in patients with a personal or family history of breast and ovarian cancer who have a "negative" 23&Me result. A negative result does <u>not</u> mean that they are not at risk. Those patients should still consider testing for other variants and other genes associated with HBOC. And as mentioned in Chapter 3, a negative genetic or genomic test does not negate the risk conferred by having a strong family health history!
- Further, all genetic testing conducted by DTC companies should be confirmed in a clinical laboratory; however, some patients (and some providers!) do not realize that confirmatory testing is required.
- To further complicate matters, there has been inconsistent messaging about DTC genetic testing from agencies such as the Food & Drug Administration. For example, in October 2018, the Food & Drug Administration authorized 23&Me to offer testing of genetic variants associated with medication metabolism (also known as pharmacogenetic testing),[10] then issued a warning against using genetic tests to predict patient response to specific medications.[11]

With all the confusion around DTC testing, its unclear implications on patient health, the risk of unexpected findings, and the sheer number of these tests being ordered by consumers, it is no surprise that patients and other health professionals are turning to genetic counselors for guidance.

Expansion of the genetic counseling field

Unfortunately, the increased utilization of and public interest in genetics and genomics is outpacing the availability of health professionals with specialized training in genetics, including genetic counselors. As mentioned earlier, the limited number of clinical genetic counselors and their accessibility outside of large health systems has resulted in long wait times for those who can find a genetic counselor, and in many cases long drives for patients residing outside metropolitan areas. In some cases, non-genetics specialists have resorted to ordering and interpreting genetic tests without consultation with a genetics professional. This is strongly discouraged and is highlighted by the 2017 lawsuit initiated after a non-genetics professional misinterpreted a genetic test, resulting in their patient undergoing an unnecessary mastectomy and hysterectomy.[12] Leaders in the genetic counseling field have recognized this growing mismatch between supply and demand, and have initiated efforts to increase the number of genetic counseling training programs. As of Spring 2019, there are 49 genetic counseling programs in the US and four in Canada. Of those, 15 are new programs established in the last few years; and three additional programs are currently under development.[13]

New models of genetic counseling

Along with the efforts to increase the genetic counseling workforce, **genetic counselors and policy makers are considering new ways to provide genetic counseling services that will meet the growing needs of patients and their referring providers**. New models that move away from the previously described traditional two-visit approach (pre-test counseling and post-test counseling) are being explored.[14] In fact, some genetic counselors have already adopted these new service models. Examples include:

- Telemedicine. Telemedicine has been utilized for quite a few years, particularly for follow-up or low risk genetic counseling sessions.[15] In addition, the Veterans Affairs Health System has been providing tele-counseling nationwide from their Las Vegas based genetic counseling program for many years. Support for this model has been growing and a number of telehealth companies have emerged in recent years with the specific goal of providing tele-genetic counseling outside of the traditional healthcare setting.
- Allied providers. "Genetic counseling extenders" including genetic counseling assistants increase efficiency by providing administrative assistance, collecting family health histories, constructing pedigrees, calling patients to report negative results, and more.[16]
- Automation. One healthcare system is even using a "chatbot" to facilitate cascade testing and share general educational information with patients and their families[17] (a chatbot is a type of artificial intelligence software that simulates an online conversation—if you've ever used the Ask Me function on a website, you're probably "talking" to a chatbot rather than a person).
- Group visits. In the United Kingdom, group visits to relay basic education about genetic and genomic tests, types of results, and expectations for genetic counseling have been used to increase efficiency and decrease the amount of time needed for one-on-one pre-test counseling with the genetic counselor.

These are just a few of the models being actively explored. Another source of innovation around genetic counseling will be the All of Us program. This study will enroll 1 million Americans in the largest cohort study ever attempted in the United States. Enormous amounts of disparate data are being collected, including genomic data. The challenges of informing participants about genomic tests (pre-test counseling) and returning results (post-test counseling) for 1 million individuals is sure to provide an opportunity for further innovation, and it may serve as a test bed for new models. As a profession, we are fortunate that the genetic counseling community has acknowledged their current limitations and are strongly committed to developing novel means of providing services to address the current barriers to timely and universal access.

Improving patient experiences with genetics

Healthcare providers, including primary care providers working with genetic counselors, can improve patient experiences in delivery of genetics services. The limited access to genetics services as well as lack of awareness of the purpose of genetic counseling have been associated with socioeconomic factors, and is greater in racial and ethnic minorities.[18] Referrals from healthcare providers have a strong influence on the likelihood of a patient attending genetic counseling.[19] However, as few as half of patients at risk for hereditary breast cancer are being referred.[20]

- Limited provider knowledge about genetics is a perceived barrier to referral and uptake of genetic counseling. Therefore, it is important for non-genetics providers to have a basic knowledge of genetics, know how to identify "red flags" in patient and family histories, and know how and when to refer to genetic counseling. Congratulations—you are overcoming this barrier by reading this book!
- It is also important that patients understand the purpose of a referral to genetic counseling so that they will actually go to their appointment. Providers can help overcome this barrier by providing an accurate description of genetic counseling (which hopefully you'll be able to do after reading this chapter) and emphasizing that genetic counseling is not the same as genetic testing. Even if patients are not interested in testing, genetic counseling can be informative and helpful in understanding their diagnosis, their disease risk and the impact on their family's health.
- Finally, providers can improve patient experiences by acknowledging their own limitations and seeking guidance through consultation or referral. As has been described throughout the chapter, genetic counselors are the genetics gurus and are here to help! Providers should not feel the responsibility to take on the practice of genetic counseling or the ordering or interpretation of genetic testing if they aren't prepared to do so. As previously described, this can have disastrous outcomes for patients (and legal implications for providers).

Summary and conclusions

In closing, genetic counselors are a wonderful resources for patients and providers. As allied health professionals specially trained in the intricacies of genetics as well as the delicate nature of patient communication, genetic counselors can ensure appropriate delivery of genetics services, improve patient outcomes, and provided necessary support to providers like you who are interested in or who have questions about genetics. To learn more about genetic counseling or make a referral, visit AboutGeneticCounselors.com/.

Rachel A. Mills is the lead author for this chapter. First person statements are from her point of view.

References

1. Baig SS, Strong M, Rosser E, et al. 22 years of predictive testing for Huntington's disease: the experience of the UK Huntington's prediction consortium. *Eur J Hum Genet.* 2016;24(10):1396–1402. https://doi.org/10.1038/ejhg.2016.36.

2. Haidle JL, Sternen DL, Dickerson JA, et al. Genetic counselors save costs across the genetic testing spectrum. *Am J Manag Care.* 2017;23(10):SP428–SP430.

3. Phillips KA, Deverka PA, Hooker GW, Douglas MP. Genetic test availability and spending: where are we now? Where are we going? *Health Aff (Millwood).* 2018;37(5):710–716. https://doi.org/10.1377/hlthaff.2017.1427.

4. Lee J. Cigna requires genetic counseling. New policy aims to reduce inappropriate testing of at-risk patients. *Mod Healthc.* 2013;43(30):4.

5. Uhlmann WR, Schuette JL, Yashar BM. *A Guide to Genetic Counseling.* 2nd ed. Hoboken, NJ: John Wiley & Sons, Inc.; 2011.

6. McCarthy Veach P, Leroy B, Callanan NP. *Facilitating the Genetic Counseling Process: Practice-Based Skills.* 2nd ed. New York: Springer International Publishing; 2018.

7. National Society of Genetic Counselors. *2018 Professional Status Survey: Work Environment.* http://www.nsgc.org; 2018.

8. Regalado A. *2017 was the year consumer DNA testing blew up.* Available from: https://www.technologyreview.com/s/610233/2017-was-the-year-consumer-dna-testing-blew-up/; 2018.

9. U.S. Food and Drug Administration. *FDA authorizes, with special controls, direct-to-consumer test that reports three mutations in the BRCA breast cancer genes.* Available from: https://www.fda.gov/news-events/press-announcements/fda-authorizes-special-controls-direct-consumer-test-reports-three-mutations-brca-breast-cancer; 2018.

10. U.S. Food and Drug Administration. *FDA authorizes first direct-to-consumer test for detecting genetic variants that may be associated with medication metabolism.* Available from: https://www.fda.gov/news-events/press-announcements/fda-authorizes-first-direct-consumer-test-detecting-genetic-variants-may-be-associated-medication; 2018.

11. U.S. Food and Drug Administration. *The FDA Warns Against the Use of Many Genetic Tests with Unapproved Claims to Predict Patient Response to Specific Medications: FDA Safety Communication.* Available from: https://www.fda.gov/medical-devices/safety-communications/fda-warns-against-use-many-genetic-tests-unapproved-claims-predict-patient-response-specific; 2018.

12. Ray T. *Oregon Lawsuit Highlights Importance of Genetic Counseling During Period of Increasing Test Access.* Available from: https://www.genomeweb.com/cancer/oregon-lawsuit-highlights-importance-genetic-counseling-during-period-increasing-test-access#.XNTz-545Kg2w; 2017.

13. Accreditation Council for Genetic Counseling. *Accredited Programs*; 2019.

14. Stoll K, Kubendran S, Cohen SA. The past, present and future of service delivery in genetic counseling: keeping up in the era of precision medicine. *Am J Med Genet C Semin Med Genet.* 2018;178(1):24–37. https://doi.org/10.1002/ajmg.c.31602.

15. Hilgart JS, Hayward JA, Coles B, Iredale R. Telegenetics: a systematic review of telemedicine in genetics services. *Genet Med.* 2012;14(9):765–776. https://doi.org/10.1038/gim.2012.40.

16. Pirzadeh-Miller S, Robinson LS, Read P, Ross TS. Genetic counseling assistants: an integral piece of the evolving genetic counseling service delivery model. *J Genet Couns.* 2017;26(4):716–727. https://doi.org/10.1007/s10897-016-0039-6.

17. Ray T. *Geisinger Deploys 'Gia' Chatbot to Help Genetic Counselors Manage MyCode Participants.* GenomeWeb; 2018. Available from: https://www.genomeweb.com/informatics/geisinger-deploys-gia-chatbot-help-genetic-counselors-manage-mycode-participants#.XNTtao5Kg2w.

18. Forman AD, Hall MJ. Influence of race/ethnicity on genetic counseling and testing for hereditary breast and ovarian cancer. *Breast J.* 2009;15(Suppl 1):S56–S62. https://doi.org/10.1111/j.1524-4741.2009.00798.x.

19. Cragun D, Bonner D, Kim J, et al. Factors associated with genetic counseling and BRCA testing in a population-based sample of young Black women with breast cancer. *Breast Cancer Res Treat.* 2015;151(1):169–176. https://doi.org/10.1007/s10549-015-3374-7.

20. Wood ME, Kadlubek P, Pham TH, et al. Quality of cancer family history and referral for genetic counseling and testing among oncology practices: a pilot test of quality measures as part of the American Society of Clinical Oncology Quality Oncology Practice Initiative. *J Clin Oncol.* 2014;32(8):824–829. https://doi.org/10.1200/JCO.2013.51.4661.

Chapter 9

Current and future trends in diagnostics and treatment

Lori A. Orlando, Brian H. Shirts, and Vincent C. Henrich

- Nongenetic (epigenetic) mechanisms can change gene activity, but have resulted in few medical applications so far.
- Many barriers impede the integration of family health history into electronic health records (EHR).
- New EHR standards and applications may facilitate integration of family health history into EHR records.
- Polygenic risk scores using common genetic variants may have moderate utility in risk stratification.
- Polygenic risk scores probably will not replace family health history for risk prediction in the foreseeable future.
- Third-party genomic direct to consumer reports based on SNP analysis often lack medical utility and may generate false positive conclusions.
- The clinical utility of genomic sequencing for adult onset disease has yet to be proven, although it has shown utility for rare pediatric diseases.

In this chapter we will describe cutting edge diagnostics and treatment options that you can expect to see and hear more about in the coming years. These are broken down into the following categories: epigenetics, digital technology, polygenic risk scores, genome arrays, and genomic sequencing.

Epigenetics

The discussions in previous chapters have considered the interaction between genetic variants, between gene variants and physiological events in the body, and between variants and environmental factors. Genetic risk is likely to be shared among relatives since family members share a significant fraction of their genetic variants, and many also share the environment in which the live and work. In fact, there is considerable evidence that genetic response to environmental exposures elevates risk for a variety of diseases, and thus families are most likely to exhibit this relationship. It has been postulated that sometimes, this increase occurs via epigenetic mechanisms. The potential

Managing Health in the Genomic Era. https://doi.org/10.1016/B978-0-12-816015-2.00009-2
181

importance of epigenetic influences for understanding familial risk is unclear. While epigenetic mechanisms have been connected to disease states, the cellular and environmental factors that trigger epigenetic processes are still largely undefined. Nevertheless, to the extent that they are triggered by environmental influences that might be shared among family members, it seems feasible that some familial disease patterns can be linked to epigenetic processes.

Clear cut environmental triggers have not been identified for most epigenetic processes

At least three types of epigenetic mechanisms have been associated with disease onset. The one which has been most widely described is gene methylation. Briefly, this is a process by which a methyl moiety is added to a CG (cytosine-guanine) dinucleotide pair in the DNA sequence. In one study, chromosomal methylation patterns in 3-year-old identical twins were found to be dramatically altered in 50-year-old identical twins by regions of both hypermethylation and hypomethylation (Fig. 1) [1], presumably resulting from environmental factors over the course of decades. Methylation requires the action of a methyltransferase enzyme, and sets off a series of changes in chromosomal scaffolding that prevents RNA transcription in the vicinity of the methylation event, and thereby silences the gene—no RNA, thus no protein, and ultimately, no protein function in the cell. Importantly, methylation does not affect the base composition of the DNA sequence (i.e., it does not involve a genetic variant). As was described earlier in the book, a cell that loses function in both copies of genes that control cell growth (e.g., *BRCA1* and *MSH2*), becomes cancerous. Instances have been observed in which a pathogenic *BRCA1* variant coupled with a normal *BRCA1* gene silenced by methylation results in cancer onset. Thus if both copies become methylated (and thus silenced), a sporadic tumor can arise in an individual without a hereditary predisposition. Further, the methylation of a few genes are known to persist across generations and exert serious effects, suggesting the possibility of an environmentally induced phenotypic change that is not dependent on inheriting a specific DNA sequence change [2].

Another non-genetic mechanism implicated in a variety of health conditions involves a family of short "noncoding" RNAs, many of which have been connected to several diseases and studied as biomarkers (to indicate the risk or onset of an adverse health event) [3,4]. Each noncoding RNA employs specific molecular mechanisms to alter cellular biology. Initially microRNAs were presumed to be junky artifacts and remnants of gene transcription. Most short microRNAs were derived from introns and intergenic regions, or from the noncoding complementary DNA strands. MicroRNAs do not possess the features of a typical gene, like protein coding sequences, or other sequence motifs associated with gene functions. In fact, the presence of microRNAs undermines the view that genes are elegant compositions with well-defined structures and functions that have evolved, acquiring their exquisite and marvelous functions through natural selection. To date, microRNA functions and processes have only been partially

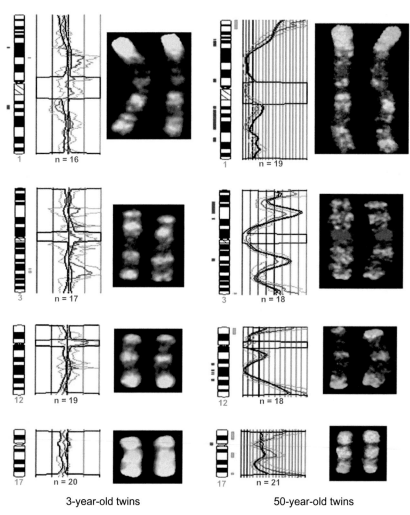

3-year-old twins   50-year-old twins

FIG. 1 Mapping chromosomal regions with differential DNA methylation in identical (monozygotic, MZ) twins by using comparative genomic hybridization for methylated DNA. Competitive hybridization onto normal metaphase chromosomes generated from 3- and 50-year-old twin pairs. Examples of the hybridization of chromosomes 1, 3, 12, and 17 are displayed. The 50-year-old twin pair shows abundant changes in the pattern of DNA methylation observed by the presence of *green* and *red* signals that indicate hypermethylation and hypomethylation events, whereas the 3-year-old twins have a very similar distribution of DNA methylation indicated by the presence of the *yellow color* obtained by equal amounts of the *green* and *red* dyes. Significant DNA methylation changes are indicated as *thick red* and *green blocks* in the ideograms. *(From Fraga MF, Ballestar E, Paz MF, et al. Epigenetic differences arise during the lifetime of monozygotic twins. Proc Natl Acad Sci U S A. 2005 Jul 26;102(30):10604–10609. Epub 2005 Jul 11. PMID: 16009939. Copyright (2005) National Academy of Sciences, U.S.A.)*

defined and are too diverse and complex to explain in depth here, but individual microRNAs can disrupt the stability of specific coding mRNAs, thereby altering a cell's protein composition, and ultimately, affecting its functional capacity. Some believe that microRNAs are a marker of ongoing inflammation and stress, processes that may underpin many of today's chronic diseases; however, the biologic pathways and triggers for increases in microRNA levels are not yet delineated clearly enough to evaluate their role as disease biomarkers [5–7]. MicroRNAs are a clear illustration of the changing perception of the genome in molecular biology. Once seen as "beads on a string," later as coding sequences of DNA, and still later as fragmented DNA sequences, the genome is now seen as a complex and somewhat disordered assembly of functional structures, that expediently employs whatever molecular entities are available to respond to environmental challenges.

A third type of epigenetic process involves the body's microbiome, that is, the community of microbe species that inhabit the body. Studies of skin, mouth, and gut microbiomes have suggested that microbial biomarkers may have utility for diagnosing imminent adverse health risks associated with inflammatory response, metabolic and nutritional disorders such as type 2 diabetes, and obesity, and neurological disorders such as Alzheimer's disease [8,9]. Further, since the microbiome is passed along from mother to offspring it can be "inherited," affecting a child's health status from birth to death, through their lifespan.

These discoveries have forced a more careful consideration of disease etiology, and beyond that, the possibility that environmental factors associated with epigenetic mechanisms are shared among family members and that the molecular players involved may be transmitted trans-generationally. Put in a more provocative way: altered molecular mechanisms may be transmitted across generations through inheritance, but are not the direct consequence of gene sequence changes. There is no denying that epigenetic mechanisms are widespread and conceivably affect health status. Nevertheless, their diagnostic value is uncertain for two reasons: First, there are very limited data on the impact of specific epigenetic events, that is, delineation of specific molecular processes with enough specificity and sensitivity to detect a disease risk or measure disease progression. It is telling that the most clear-cut examples of an environmental influence acting on gene activity come from studies related to the effects of starvation, an extreme and unmistakable environmental factor [10]. These examples are important, but may not be representative of less extreme nutritional situations. Second, while epigenetic events conceivably have value as a diagnostic tool, the pathways from trigger to specific epigenetic change to change in health status have not been delineated. This is not an indictment of epigenetic mechanisms, but rather an acknowledgment that we know a little and we don't understand a lot when it comes to analyzing precisely the molecular underpinnings of chronic disease.

Epigenetic mechanisms are not "purely" environmental

The recognition of epigenetic phenomena has rekindled the nature vs. nurture debate. Most likely, this debate will once again lead us to the recognition that there is a constant interplay of genes and environment, and that neither can be entirely ignored. A full understanding of disease susceptibility and onset depends on examining both factors. Consider methylation for instance. The process itself does not require a change in DNA sequence to occur. However, the process acts on genes that encode proteins that in turn, mediate chromatin changes imposed by methylation. The mutational process goes on in all of us, all of the time, and a new genetic variant affecting methylation response to the environment could arise anywhere in the genome—and probably would not be detected in a "big data" scan. Similarly, microRNA sequences are derived from DNA sequences, and in a few instances, microRNA SNPs have been associated with disease risk in GWAS studies [11]. Finally, while the microbiome consists of organisms that do not affect the human genome directly, individual reactions to microbes are highly variable. Novel drug targets were identified in a gene coding for T-cell receptors (CCR5) that serves as an entry point for HIV infection by discovering a common variant in the genomes of individuals seemingly immune to the virus [12]. A viral or bacterial infection is environmental, but the host response can be genetically coded and variable in a population. Influenza, is another example. It, of course, is a common human disease with infection cycles that have been historically punctuated with widespread epidemics. In an epidemiological study, a family was identified whose members lived normal uneventful lives, except during influenza epidemics. When that happened some family members showed a distinctive response—they died [13]. All viruses act by wresting control of cellular processes, and most of us mount a response that wards off the attack. In the HIV case, a genetic variant reduced susceptibility to the virus, however, in the influenza case a genetic variant in the family enhanced susceptibility by compromising the body's response to infection with disastrous consequences. Understanding how mutations increase and decrease susceptibility to infections of all types (not just viruses), can provide insights into the line between physiology and pathology and how to optimize resistance.

When contemplating the factors that lead to disease vulnerability, we repeatedly see an interplay between the factors that trigger disease progression and the body's response to them. To summarize the point of this book, a positive family health history for a chronic condition offers a convenient risk assessment for the patient even when more precise information about the factors responsible, genetic and environmental, are not easy to identify. As the inherited and shared environmental factors that lead to a positive health history are better defined, the need to develop precise diagnostic tools to evaluate the effect of these factors on health status will intensify concomitantly.

Integrating digital family health history into the EHR and consumer apps

Remember back in Chapter 2, the discussion about the clinical utility of family health history and its value in clinical care? Well, let's say you are convinced that it is time to make better use of collecting and analyzing family health history.

What's the challenge, why aren't we doing this already? The problems are multi-faceted and cross stakeholders.

- Patient-related barriers fall into two categories. (1) Disinterest. Family history is not particularly scintillating, requires a great deal of communication and effort to gather, and many don't appreciate the benefit that undergoing a family history based risk assessment can have on their future (or even current) health. Thus, the combination of effort and low perceived utility turns many away—particularly younger individuals, who would actually benefit the most from risk assessment. (2) Knowledge. This was touched on in Chapter 2, when describing the accuracy of self-reported family health history. The information is there, but not always easily accessible. It is often scattered across relatives and sometimes buried in the family archives; but if individuals know WHAT to ask about and HOW to ask they can substantially improve the likelihood of obtaining the necessary data.
- Provider-related barriers include a lack of awareness about the impact of family health history, the length of time it takes to collect a high quality family health history that is suitable for a risk assessment, and an inability to synthesize the complex data into a meaningful action plan during the time allotted for a clinical encounter. This book is meant to help address the first! We hope that our message is clear and that by now we've converted you all to family health history aficionados. But what about the other two? The first, is typically addressed using a family health history intake form prior to the appointment, moving the data collection component to pre-provider; however, the barriers described above limit the quality of the data that can be collected on the spot at a visit. The second, requires assistance. Today's algorithms tend to be logistic regression models that can't be calculated without a computer, and even those that aren't regression models, like the National Comprehensive Cancer Network's guidelines for when to test for a hereditary condition, are over 80 pages long. In addition, there are over 50 different hereditary conditions to incorporate as part of the risk assessment. Online calculators are immensely helpful but they are scattered all over the internet and the data must be manually entered. They are useful, but the process is still too time consuming for the average clinician.
- Health system barriers include failure to provide the necessary tools to overcome the barriers on the patient and physician side. Even electronic medical records, which held such promise for simplifying the process, do nothing to facilitate data collection, data storage, or data analysis.

The convergence of these barriers has stymied the potential of risk assessment to improve population health. Reviews of medical records, even electronic ones, repeatedly show that <5% of patients have an adequate family health history recorded to perform a risk assessment, much less actually perform one. If performing risk assessment on just a family health history is this complex, think how difficult it will be once genomic data starts making its way into the risk algorithms!

One solution to these complex and inter-related challenges is to leverage the burgeoning field of digital health. Patient-facing web- or computer-based health risk assessment tools could move the data collection component from the physician's office to the patient's home, perform real time data analysis, and generate clinical decision support for patients and providers at the point-of-care. Today there are several family health history based risk assessment platforms available, some that are embedded into clinical practices, and others that are direct to consumer. These platforms are all designed to promote greater patient engagement and accuracy of family health history information, but in different ways [14,15]. A listing of many of these can be found in the Global Alliance for Genomics & Health's Genomic Data Toolkit (https://www.ga4gh.org/genomic-data-toolkit/). However, to improve healthcare delivery and promote population health, a family health history based digital risk assessment platform should include the following characteristics: computerized, patient administered, easy to use, collects all data necessary for risk stratification, updateable, has integrated risk algorithms and evidence-based clinical decision support, and can communicate with the EMR [16]. In comparison the major EMR vendors are not patient administered, not easy to use, do not collect all of the necessary data, and do not have integrated risk algorithms or clinical decision support.

In the optimal scenario family health history data collection is removed from the actual clinic visit (Fig. 2). Patients, empowered and educated on how to gather high quality and thorough family history information, confer with relatives, then access a patient-facing platform where they enter their family data. After data entry, risk algorithms run in real time and provide clinical decision support to both patient (in real time) and provider (at the point of care). MeTree, a family health history based risk assessment platform, developed by Dr. Orlando and others employs this approach. It integrates with electronic medical records using a newer data standard (SMART on FHIR) that allows patients single sign on access to the platform via their patient portal, imports existing electronic medical record data needed for the risk assessment, permits logging in and out to gather data and update as needed, provides context sensitive help and embedded educational materials to enhance data collection, runs clinical decision support for 50 conditions, and generates a graphical dashboard with key points for patients and with action items and links to evidence-based guidelines for providers that can be viewed from within the patient's chart in the electronic medical record (thus not disrupting the provider's workflow). The clinical utility of digital risk assessment platforms, like MeTree, has been consistently demonstrated, with studies indicating that a much larger percentage of the general population is at high risk for hereditary conditions, 20–25%, than previously expected, and that most are not being identified as part of current practice [17].

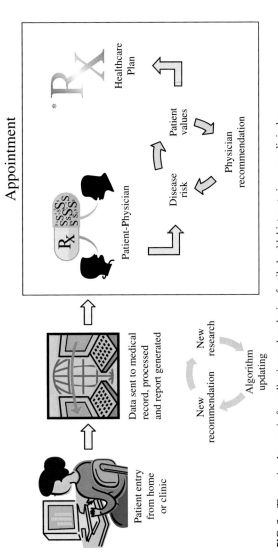

FIG. 2 The optimal scenario for collecting and analyzing family health history to improve clinical care.

Two additional benefits of a digital health platform with the characteristics outlined above, are the potential to reduce bias and to access populations that to date are not well represented in traditional healthcare systems. Bias is inherent to all humans, it is how our brains work, we develop heuristics based on our experiences and knowledge, and match our diagnostic and critical thinking to those heuristics. These have served us well as a species, but can often lead us astray in today's society with its extensive gaps in knowledge, particularly among minority and poor populations who don't typically participate in clinical trials. The benefit of computers is that, to the extent that our bias is not programmed into their algorithms, they can analyze data without referring to heuristics. Gaining access to under-represented populations, something else a digital platform may facilitate, could help fill those gaps. The two distinctly different underrepresented populations under consideration here are minority/under-served individuals and young adults. The approach to engaging each is different and requires different types of platform features. For minorities and the underserved, access is a critical feature of the platform. Not having to leave work or find transportation to a facility will lower barriers to engaging. In addition, simplicity of the interface, culturally appropriate content at a low literacy level that resonates with minorities, and enabling a complementary smartphone application, will be critical. Of course, while these features can help, they don't address issues of trust and other deeply ingrained barriers to engaging with the health system. Young adults, on the other hand, have completely different, but not mutually exclusive, challenges. By and large they take being healthy for granted and are not as concerned about the future as they are with "today." For that reason they rarely engage with the health system, except women who need birth control, or who are pregnant. Even messages raising awareness of health risks that could be averted through early risk assessment, don't have much of an impact. Those who do respond to such messaging tend to be those who are already primed, by having particularly "bad" family health histories. To facilitate engagement with these groups digital platforms will need to adopt strategies that work for other types of young adult online platforms, like gaming platforms, that include competitive tasks, visually appealing material, social features, and short modules. Successfully engaging these two populations could transform healthcare.

Polygenic risk scores

Medically relevant single nucleotide polymorphisms (SNPs) have been employed to develop polygenic risk scores for various diseases

Single nucleotide polymorphisms (SNPs) and indels (insertions and deletions) have been incorporated into panels that evaluate the risk of various forms of cancer and other diseases. For example, prostate cancer SNP panels were developed several years ago (e.g., 40), and the general approach is now widespread, and serves as the foundation for algorithmic polygenic risk scores estimating an individual's inherited risk for a growing number of chronic diseases. A brief primer is offered here to illustrate how these SNPs are identified and utilized. This type of SNP panel differs from the targeted diagnostic panels which report the presence or absence of risk-elevating variants (including the two SNPs associated with the *APOE* variants) or diagnostic sequencing panels that comprehensively sequence a panel of genes. Polygenic risk score panels are conceptually based on the notion that several genetic variants interact to increase a person's risk. Prostate cancer is a good example of a disease with potential advantages for testing with a polygenic risk score. Recall that in the Scandinavian twin study, the concordance of prostate cancer between male twins was strongly impacted by heritable factors while shared environment exerted no measurable influence on twin cancer concordance? The relatively high heritability of prostate cancer, its high incidence (it is globally the most common form of cancer in males), its relatively late onset, and the lack of any known environmental risk factors suggest the possibility that predictive genetic testing could offer a significant benefit to patients.

Over the past few decades, several genome-wide association studies (GWAS) have implicated specific SNPs as potential indicators of prostate cancer risk. Clinically, prostate cancer is highly variable, including its age of onset, its rate of progression, its connection to prostate specific antigen levels (sometimes it is predictive, sometimes not), and its association with other hereditary cancer syndromes. The SNPs for genome wide associations studies were originally selected for their relatively high frequency in the population, and not for their role in disease causation or function. In fact, most SNPs used in genome wide association studies have no functional roles in the genes involved. Instead, they tend to be co-located near the SNPs that are involved, but those SNPs are so uncommon that they are hard to detect. This process, using SNPs that are markers of disease, but not actually involved in the disease, is called linkage disequilibrium. The general principle of linkage disequilibrium is that if two SNPs travel together on the same chromosome, they will occur in connection with each other more often than is expected by chance.

How SNPs identified as a marker of disease by genome wide association studies alter protein function or disease risk is usually not known. For example, at least one prostate cancer associated SNP has also been tied to type 2 diabetes in other studies. The diversity of effects implied by different association studies could mean that some SNPs influence cellular activities that are not specifically tied to cancer cell growth, but rather to cellular functions that influence cancer onset and progression (for instance, metabolic and nutrient regulation). Because of the way SNPs are selected for panels, most statistically associated SNPs will not have a direct biological effect, but will rather be markers that are physically linked to other variants.

Generally, the diagnostic value of polygenic risk scores requires a rigorous analysis of their clinical validity and utility, including how they might impact the course of treatments, and the proportion of patients who might benefit from the test results. These issues concerning the utility and validity of prostate cancer gene panels have been discussed in recent medical reviews [18]. At this time, the larger issue of polygenic risk scores' clinical validity and utility remains unresolved though they have become an intense area of interest and investigation. These are important considerations even before contemplating the possibility that other non-SNP variants and cellular processes may be tied to an individual's disease risk.

Expanding the size and scope of polygenic risk scores—Future challenges and risks

Early polygenic risk score successes and the establishment of large research cohorts with SNP data, have enabled additional research on polygenic risk scores, particularly in other disease areas. In fact, the polygenic risk score framework is the basis for disease risk calculations generated by companies like 23andMe. However, for diseases with smaller heritable components than prostate cancer, polygenic risk scores may be more difficult to understand.

Some polygenic risk score panels now include, not just the SNPs that achieved genome wide significance, but complex calculations that incorporate data from dozens, hundreds, thousands, or even millions of less significant variants (those that did not reach the statistical threshold). The theory behind incorporating these non-significant SNPs as part of the score is that they may carry cryptic information, which enhances prediction when taken as a whole. The concern about this approach is that while SNPs associated with disease may be markers for SNPs with a molecular effect (via linkage disequilibrium), they may also be associated with non-genetic factors in the population. The largest polygenic risk score studies create models that generate risk scores three or four fold higher than average risk, on the level with those generated by high penetrance mutations or other major risk factors [19,20].

It can be difficult to understand exactly what is happening inside of a polygenic risk score. This is because no one really knows exactly what accounts for their predictive success. However, we can think of their impact using 4 large conceptual categories as a framework, though, for any given variant it is usually not clear which categories the variant falls into.

(1) SNPs that are linked to a causative variant

These variants have a meaningful association, usually of small effect size, with underlying disease biology. These are usually top hits from the ever growing genome wide association studies that have been validated as genome-wide significant in multiple large case-control trials [21]. For example, a SNP variant strongly predisposing for prostate cancer was discovered in *HOXB13*, a gene that had not previously been associated with any type of cancer. The *HOXB13* gene is a homeotic gene involved in the differentiation of specific cell types. However, while the embryonic role of *HOXB13* and homeotic genes have been well-established for decades in a variety of genetic model organisms, the role of these genes in adults remains largely unexplored [22]. As often happens with unexplained observations in cell biology, the activity of homeotic genes in post-embryonic stages has simply been overlooked, and a search to discover what the protein is doing at a different developmental time could be seen as open-ended, expensive, and futile without a prior connection to a serious and widespread adult disease. Even more important than its general relationship to prostate cancer, was the finding that it was associated with a predisposition for aggressive and malignant forms of prostate cancer, indicating its prognostic value and a potential clinical role for intensified monitoring of carriers. If only variants with known functional consequence were included in a risk score, it would theoretically be useful in anyone; however, common allele frequencies vary widely between populations so validated polygenic risk scores in one population may be less predictive in a different population. Differences are particularly common between ethnically divergent populations.

(2) Variants that are indirectly associated with other variants or other risk factors

These SNPs are often common variants that are in linkage disequilibrium with other, rarer variants that cause disease (as described earlier). There are likely to be a great many of these types of variants, and they often have low association scores. Although, the genetic associations are real, it can be challenging to understand why. Is a SNP with a small effect size predictive because it is linked to a nearby (100 kb away) low impact but common causative SNP (population prevalence of 5%) or to a very close (10 kb away) high impact, but rare (population prevalence 0.1%) causative SNP.

These may also be SNPs that are indirectly and not causally related to the disease of interest's biology and risk. For example, SNPs associated with diabetes may be indirectly associated with cardiovascular and kidney disease, because diabetes damages the heart vessels and kidneys, and is thus a leading cause of those diseases. SNPs can also be associated with environmental conditions. For example, associations with lipid levels may actually be due to a secondary pharmacogenetic response to statins in the subset of the population that is being treated for high lipid levels.

This category of polygenic risk score impact does not transfer well between ethnic populations, because marker SNPs are population specific and thus different marker SNPs may be linked to causative variants in different populations. Because of this polygenic risk scores have not transferred well across different ethnic populations. Use of these marker SNPs, rather than variants with a known molecular effect, is one limitation to the widespread deployment of polygenic risk scores to predict risk in the general population. Currently most polygenic risk scores have only been tested in populations of Northern European ancestry. Thus, there is a substantial risk that without accounting for ethnic diversity, broad clinical use of current polygenic risk scores could exacerbate health disparities.

(3) Variants that are cryptically associated with social factors or other correlates of disease

These variants do not reach robust statistical significance and are less likely to be replicated across genome wide association study analyses in different populations [23]. This has been most clearly illustrated in Finland, where polygenic risk scores for several diseases have been linked with social and geographic clines, rather than genetic causality [24]. Income and socioeconomic status may be correlated with country of ancestral origin or ancestral pedigree within a country. Local geography is also known to correlate with genetics, in subtle ways, and with disease risk. For example, fishers may have different diets than farmers, and thus may have different disease risks. Though occupations are not genetic, trades tend to be passed down in families, and thus could be distinguishable on SNP panels. Educational level is highly correlated with many diseases, and also passed down socially, creating a correlation between pedigrees and non-causal genetic variants. This may be another major reason why polygenic risk scores do not transfer from one ethnic population to the next. If a polygenic risk score is generated and validated in two different cohorts, but both are from the same population, it is very likely that there will be cryptic genetic associations.

Two problems with these cryptic associations are that they are population specific and that there is no reason to expect predictive validity going forward. Since cryptic associations are based on population specific social and geographic trends, polygenic risk scores based on cryptic association will only be valid for patients in that population. Because cryptic associations are inherently based on genetic markers reflecting how historical demographic trends affect families, they may replicate well with multiple retrospective samples, but fail in prospective predictions as demographic trends change. Because of these problems, it is challenging to envision a polygenic risk score derived from a large number of SNPs that could be clinically useful or that would not exacerbate healthcare disparities.

(4) Overfitting

If a polygenic risk score model has not been validated in a separate, independent population, it will be statistically overfitted. The concept of overfitting means that the model works perfectly in the specific population that is created for, but is so closely tuned to the idiosyncrasies of that population that it does not work well in any other group. Overfit SNPs distinguish cases from controls no better than chance. Overfitting is a common problem with statistical models built from large datasets (so called big data). The best polygenic risk score studies validate the polygenic risk scores in completely separate test and validation datasets, preferably from two completely different populations. If a polygenic risk score has not been rigorously validated using this process, then it is not valid retrospectively or prospectively.

Polygenic risk scores and family history in clinical care

Do not feel bad if you still do not understand polygenic risk scores. You are not alone! Until these categories are sorted out, no one will really understand exactly how polygenic risk scores work; not even those who developed them.

There are growing, and increasingly successful, efforts to understand category 1 SNPs (https://www.ebi.ac.uk/gwas/), and a few early efforts to sort out those in category 2. Category 2 is more challenging as it can be difficult to identify all the medical correlates for any given disease, which is necessary, before disentangling them from genetic associations. Category 3 is less of an issue for polygenic risk scores that use a few dozen or even a few hundred variants; but it is a major challenge for polygenic risk scores that use thousands of variants. Category 4 is a result of lazy science, and almost all polygenic risk scores published in high quality journals address overfitting. However, there is still a high risk that someone will claim a proprietary polygenic risk score has amazing predictive value when the ultimate result is due to overfitting. This will be particularly problematic for scores derived by companies who will not publish their results in order to protect their intellectual property. Overfitting can be difficult to identify for those with little experience in big data genetic analyses.

One large study that recently compared a polygenic risk score with family history found that 10–20% of the information contained in the family history was also in polygenic risk score [20]. This means that even with the best polygenic risk score over 80% of the information gained from family health history is independent and not captured by a polygenic risk score. Given these challenges, polygenic risk scores will most usefully be applied within the well-defined populations from which they were derived and (hopefully) validated. In addition, they may not discriminate risk at the individual and family level as well as they do at the population level. Therefore, it is unlikely that polygenic risk scores will replace family health history for risk prediction, though polygenic risk scores may be used to supplement it.

Once again, for the practitioner who cannot select which patients to manage, a positive family health history offers a relatively straightforward and economical strategy for assessing a patient's future disease risk and identifying an appropriate screening and prevention action plan to prevent, or at least detect disease onset early. Could a polygenic risk score lead to robust predictions about which members of a family are carriers of a "genetic load" that leaves them at greater risk for a given chronic disease than other members? There is little evidence for or against this assertion yet. The SNPs used in polygenic risk scores represent only a fraction of genetic variation, and only variants that are common in the population. This fact further highlights the value of family health information when assessing disease risk, especially when genetic factors are suspected. Thus, family health history currently provides a more reliable and cost-effective strategy for medical decision making than a polygenic risk score.

Genome-wide SNP arrays

Genotyping and sequencing data are becoming increasingly common as costs of obtaining these tests continue to drop. At the time of this writing, over 25 million people have ordered direct-to-consumer SNP-based genotypes for genealogy and other trait information through commercial companies like AncestryDNA, MyHeritageDNA, and 23andMe. These services link individuals to distant relatives and help identify common ancestors. Most direct-to-consumer (DTC) genetic testing services do not report health related variants, with the exception of 23andMe, which has a complicated history. Initially, its business model focused on ascertaining genetic health, but pivoted to ancestry and non-medical traits after receiving a restraining order from the Food & Drug Administration. Recently, with guidance from the Food & Drug Administration it is pivoting back to health-related risk results.

The genotyping performed for all these tests are done using the same SNP arrays developed during research studies. The main difference between the different types of genetic testing is not the SNPs that are tested. Commonly used SNP arrays are comprised of variants known to be common in the population, and widely accepted by the community. The core difference between genomic SNP arrays used for research, 23andMe, AncestryDNA, and clinical cytogenetics is in the analysis and, most importantly, the interpretation of the SNP data. In fact, there have been quite a few studies showing that different companies interpret the same results differently, depending upon their understanding of the literature.

To address variations in interpretation third-party analytic companies have arisen offering second opinions on existing genotype results. If you have done testing using AncestryDNA, 23andMe or similar service, you can download your SNP calls today, and submit them online to one of these companies. Third-party analysis companies provide a second opinion about your ancestry and/or important traits. They may even offer to analyze your data and return a "clinical" report. The reports almost always include disclaimers that results should not be used for medical care, even though they are marked as clinical reports and written with disease related interpretations and medical suggestions. There are currently no third-party genomic reports that are validated by CLIA compliant laboratories or approved by regulatory agencies. At many institutions a substantial proportion of genetics consultations are to evaluate reports from direct-to-consumer testing or third-party analysis of such testing. Unfortunately, these consultations are almost always fruitless. Findings are usually not medically actionable, especially in healthy adults. There have also been concerns about analytic validity—the reproducibility of the finding. For example, if one laboratory finds a pathogenic variant at a specific location in a gene, a different lab should also find the same variant in the same location. This, however, is not always the case and there have been numerous examples of a pathogenic variant identified by one company that is not found when the sample is re-run in a CLIA compliant clinical laboratory. Because direct to consumer testing is becoming more common, providers should be prepared for patients to ask about it. An increasingly common response is to ascertain if there is a specific concern, address the concern, and redirect the patient to the principal purpose of the visit in order to avoid expending the bulk of their increasingly short visits on medically questionable results. Patients may be upset with providers who dismiss direct-to-consumer genetic test results out of hand, so providers should be careful about tone and the wording of their responses to these queries.

Genomic sequencing

If you wanted to, you could also sequence your own genome for a few hundred dollars today. Although sequencing is relatively inexpensive, quality interpretation is more challenging. There are so many more variants to take into consideration with sequencing, and the vast majority of them are in introns (areas of the gene that are excised prior to encoding a protein), intergenic regions, or in noncoding regions of exons. The enormous amount of data contained in a single genome is overwhelming and storage alone is a challenge, much less analyzing it in a meaningful way. Everyone has variants that are unique to themselves and their families. Most personal and family specific variants do not cause disease or increase disease risk. Determining the clinical significance of a family specific variant can be extremely challenging. Currently, the greatest successes in genomic sequencing have been in understanding rare syndromic pediatric diseases. In the context of pediatric disease one of the most powerful strategies is sequencing the child and both parents. This allows clinicians to determine which findings are unique to and thus de novo in the child. Evaluating unique syndromic features against a few de novo mutations often highlights one that is by far the most likely candidate. Parents that are sequenced for the purpose of interpreting the sequencing results in children with rare pediatric syndromes may be offered the option of receiving medically actionable results. These "incidental findings" reports are usually limited to the sequencing results of 59 genes deemed medically actionable by the American College of Medical Genetics and Genomics. Genes on this list include *BRCA1*, *BRCA2*, *MYH7*, Lynch Syndrome genes, and many of the other genes listed in this book.

Unfortunately, determining the functional consequences of variants in genes not on the short list of genes with established medical actionability in adults is more challenging. Particularly as the longer we live the more variants we acquire and if the pathologic variant was one we were born with or acquired? If acquired it will be present in some but not all cells, which is why we sequence tumors- to understand the driving mutations of the tumor which would not be present in the baseline genome. More sequencing will undoubtedly lead to cataloguing and characterizing more familial variants. Family health history will continue to be an important tool in understanding genomic sequencing results.

All three authors contributed to the content of this chapter.

References

1. Fraga MF, Ballestar E, Paz MF, et al. Epigenetic differences arise during the lifetime of monozygotic twins. *Proc Natl Acad Sci U S A*. 2005;102(30):10604–10609. https://doi.org/10.1073/pnas.0500398102.

2. Ferguson-Smith AC. Genomic imprinting: the emergence of an epigenetic paradigm. *Nat Rev Genet*. 2011;12(8):565–575. https://doi.org/10.1038/nrg3032.

3. Boon RA, Dimmeler S. MicroRNAs in myocardial infarction. *Nat Rev Cardiol*. 2015; 12(3):135–142. https://doi.org/10.1038/nrcardio.2014.207.

4. Petrovic N, Davidovic R, Bajic V, Obradovic M, Isenovic RE. MicroRNA in breast cancer: the association with BRCA1/2. *Cancer Biomark*. 2017;19(2):119–128. https://doi.org/10.3233/CBM-160319.

5. Arner P, Kulyte A. MicroRNA regulatory networks in human adipose tissue and obesity. *Nat Rev Endocrinol*. 2015;11(5):276–288. https://doi.org/10.1038/nrendo.2015.25.

6. Hollins SL, Cairns MJ. MicroRNA: small RNA mediators of the brains genomic response to environmental stress. *Prog Neurobiol*. 2016;143:61–81. https://doi.org/10.1016/j.pneurobio.2016.06.005.

7. Kumar S, Reddy PH. Are circulating microRNAs peripheral biomarkers for Alzheimer's disease? *Biochim Biophys Acta*. 2016;1862(9):1617–1627. https://doi.org/10.1016/j.bbadis.2016.06.001.

8. Baothman OA, Zamzami MA, Taher I, Abubaker J, Abu-Farha M. The role of gut microbiota in the development of obesity and diabetes. *Lipids Health Dis*. 2016;15:108. https://doi.org/10.1186/s12944-016-0278-4.

9. Buford TW. (Dis)trust your gut: the gut microbiome in age-related inflammation, health, and disease. *Microbiome*. 2017;5(1):80. https://doi.org/10.1186/s40168-017-0296-0.

10. Tobi EW, Goeman JJ, Monajemi R, et al. DNA methylation signatures link prenatal famine exposure to growth and metabolism. *Nat Commun*. 2014;5:5592. https://doi.org/10.1038/ncomms6592.

11. Moszynska A, Gebert M, Collawn JF, Bartoszewski R. SNPs in microRNA target sites and their potential role in human disease. *Open Biol*. 2017;7(4). https://doi.org/10.1098/rsob.170019.

12. Becker Y. The molecular mechanism of human resistance to HIV-1 infection in persistently infected individuals – a review, hypothesis and implications. *Virus Genes*. 2005;31(1):113–119. https://doi.org/10.1007/s11262-005-2503-5.

13. Albright FS, Orlando P, Pavia AT, Jackson GG, Cannon Albright LA. Evidence for a heritable predisposition to death due to influenza. *J Infect Dis*. 2008;197(1):18–24. https://doi.org/10.1086/524064.

14. Cleophat JE, Nabi H, Pelletier S, Bouchard K, Dorval M. What characterizes cancer family history collection tools? A critical literature review. *Curr Oncol*. 2018;25(4):e335–e350. https://doi.org/10.3747/co.25.4042.

15. Welch BM, Wiley K, Pflieger L, et al. Review and comparison of electronic patient-facing family health history tools. *J Genet Couns*. 2018. https://doi.org/10.1007/s10897-018-0235-7.

16. de Hoog CLMM, Portegijs PJM, Stoffers HEJH. Family history tools for primary care are not ready yet to be implemented. A systematic review. *Eur J Gen Pract*. 2014;20(2):125–133.

17. Orlando LA, Wu RR, Myers RA, et al. Clinical utility of a web-enabled risk-assessment and clinical decision support program. *Genet Med*. 2016;18:1020–1028. https://doi.org/10.1038/gim.2015.210.

18. Little J, Wilson B, Carter R, et al. Multigene panels in prostate cancer risk assessment: a systematic review. *Genet Med*. 2016;18(6):535–544. https://doi.org/10.1038/gim.2015.125.

19. Khera AV, Chaffin M, Aragam KG, et al. Genome-wide polygenic scores for common diseases identify individuals with risk equivalent to monogenic mutations. *Nat Genet*. 2018;50(9):1219–1224. https://doi.org/10.1038/s41588-018-0183-z.

20. Mavaddat N, Michailidou K, Dennis J, et al. Polygenic risk scores for prediction of breast cancer and breast cancer subtypes. *Am J Hum Genet*. 2019;104(1):21–34. https://doi.org/10.1016/j.ajhg.2018.11.002.

21. Mills MC, Rahal C. A scientometric review of genome-wide association studies. *Commun Biol*. 2019;2:9. https://doi.org/10.1038/s42003-018-0261-x.

22. Ewing CM, Ray AM, Lange EM, et al. Germline mutations in HOXB13 and prostate-cancer risk. *N Engl J Med*. 2012;366(2):141–149. https://doi.org/10.1056/NEJMoa1110000.

23. Berg JJ, Harpak A, Sinnott-Armstrong N, et al. Reduced signal for polygenic adaptation of height in UK Biobank. *Elife*. 2019;8:e39725. https://doi.org/10.7554/eLife.39725.

24. Kerminen S, Martin AR, Koskela J, et al. Geographic variation and bias in the polygenic scores of complex diseases and traits in Finland. *Am J Hum Genet*. 2019;104(6):1169–1181. https://doi.org/10.1016/j.ajhg.2019.05.001.

Chapter 10

Current and future trends to integrate family health history with clinical programs to improve population health

Brian H. Shirts, Lori A. Orlando, and Vincent C. Henrich

- All humans are genetically related as part of one large, extended family.
- Technology allowing health systems to document and use family health history more efficiently and effectively has high potential to improve medical care.
- As genetics increasingly becomes an integral part of medical care, there will be more genetics providers and better tools to help non-geneticists incorporate genetics into standard care.
- Other resources, such as Facebook and online genealogy databases, are increasingly used to connect large extended family groups in efforts to trace inherited genetic diseases.

We are all related

When talking about recessive disease in my undergraduate course on hereditary disease I like to ask for a show of hands from all of those who have a loop in their family tree. This is typically called inbreeding. Very few students raise their hands, but I raise my hand. I tell the students that I know that I am my own 11th cousin twice removed. I also found out that I married my 9th cousin once removed. I then tell the students that they should all raise their hands, because everyone has loops in their family tree, most people just don't know about them.

Choose two random individuals from the same broad ethnic background and they are likely to have at least one common ancestor 12 generations back. One thousand years ago, back in history, those two will have more common ancestors than unique ancestors. Theoretically, there are ancestors 3000–5000 years ago that can claim the entire human family as their descendants. Our families extend beyond our close relatives to our local community and ultimately to the entire world.

Managing Health in the Genomic Era. https://doi.org/10.1016/B978-0-12-816015-2.00010-9

This final chapter will discuss ways that family health history can extend beyond the typical 3-generation pedigree to distant relatives and to the whole human family.

Healthcare models that promote health through new processes

As the complexity of family health history, genetics, and genomics continues to increase, the number of guidelines that are based on risk or genetic/genomic markers expand, and healthcare encompasses new value-based care models that disincentivize fee-for-service clinical care, new strategies for delivering care that also maintain health are essential. As the focus of primary medical care shifts to health maintenance, the need for broader implementation of family health history and genetics/genomics will drive novel strategies and tools for patient management, but will also create a host of concerns that need to be addressed. Three strategies that have enormous potential to be both cost effective and health promoting are optimizing data collection, systematic risk assessment, and extending the reach of genetic counseling.

In Chapter 9, digital family health history based risk assessment platforms were described. To use these platforms in new "health" based medical systems, the platforms need to expand beyond just the essential characteristics necessary to be an effective tool (computerized, patient administered, easy to use, collects all data necessary for risk stratification, updateable, has integrated risk algorithms and evidence-based clinical decision support, and can communicate with the electronic medical record), to encompass strategies that facilitate optimal data collection. Particularly since the utility of a platform hinges upon data quality and quantity. Some strategies, such as embedding education and importing personal medical history into the platform from the medical record, were previously mentioned; however, going beyond these, we could consider incorporating strategies that leverage social networking to facilitate collaboration among relatives and that link each individual relative to their electronic medical record in order to import their stored personal health information into the pedigree. The two are complementary strategies, as smaller practices are unlikely to transition to an electronic medical record and not all electronic medical records support data sharing. In terms of social networking a number of opportunities exist: Facebook like permissions-based models that link relatives in a family network together to discuss medical conditions and enter their own or others' information into the digital platform, virtual reality environments that allow family members to gather together in space/time and collaboratively build and reach consensus on medical history, or better yet augmented reality which doesn't rely on expensive devices and thus expands the reach to the underserved. One step beyond relatives working together to build their family medical history, is for each relative to link to the digital platform through a SMART on FHIR interface that imports their personal health information and then links all the relatives together in a single large family "genealogy." The benefit of this model is that the electronic medical record often contains substantially more detailed information about an individual's personal medical history, for example tumor pathology, than can be collected via self report, and it significantly reduces the amount of effort needed to gather the family history. The downside is that family members must trust each other enough to create a direct link to their medical record and to share that information with others. This active agreement to sharing, has a higher bar than the passive transfer of information about one family member to others.

Going one step beyond the platform itself, a systematic risk assessment could be integrated into health systems as an entirely separate pathway that does not rely on overwhelmed primary care providers. As new technologies and healthcare processes come to light they, by and large, fall into the lap of the primary care provider to deploy and manage. Yet, they are among the lowest paid providers, forcing them to have high throughput clinics to make ends meet. They rarely have the luxury of time to take on these new tasks, or to undergo the "education" needed to appropriately deploy them. Rather than continuing down this path, a new health system risk assessment strategy could, for once, offload some of this effort. This pathway would start with all patients coming into a health system (and existing patients if they have never undergone a risk assessment) accessing and completing an integrated digital risk assessment platform. A healthcare provider (nurse, physician, physician extender, etc.) would review the results and contact the patient for a consultation, if any risks are identified. The recommended risk management strategy, whether it is breast MRI or genetic testing, would be reviewed, and the key data elements triggering the recommendation, confirmed. After a shared decision making conversation, if agreeable, the patient would undergo the recommended strategy, and, if it includes genetic testing, an individual with expertise in genetics (trained nurse, genetic counselor, etc.) would explain the process and return results (or if a hereditary clinic exists in the health system they could be scheduled into that clinic). If positive for a hereditary syndrome the patient would be scheduled into the appropriate specialty clinics, and if not (or if they were at familial risk, and only needed an MRI for example) they would be scheduled with the primary care provider along with documentation of their risk and the recommended management. In addition, the process would be iterative, so that whenever there was a change in their family history, a patient would update the platform. If new risks were identified, they would be contacted and the same process followed. In this way providers could develop experiential learning as patients are returned to them with information about their risk and appropriate management, without increasing their workload.

In addition, the pathway could utilize a small team with the skill set necessary to efficiently and effectively implement systematic risk assessment, which will in turn identify a significant number of individuals who meet enhanced risk screening and would not otherwise have been diagnosed. Notably, if this process is linked to a digital platform that enables relatives to connect their medical records, each connected health system benefits every time an individual's diagnoses are updated. An added benefit is that if a relative is diagnosed with a hereditary condition, it immediately facilitates cascade screening in a scalable and efficient manner.

As described in Chapter 8, genetic counseling is currently a rate limiting step in the clinical implementation of genetics and genomics. And this is based on today's demand. Just think of what it will be like if health systems start performing systematic risk assessment. The demand is likely to triple or even quadruple in a few short years. Unfortunately, the number of genetic counselors trained each year is small, the capacity of the training programs is limited, and the lure of genetic testing companies has kept the number of practicing genetic counselors well below the demand. In addition, they tend to be located in large academic medical practices where the number of individuals needing their services is high. The same economic forces come to bear for many specialties, where there is an overabundance in densely populated metropolitan areas, but significant scarcity outside these areas. Solutions to this particular challenge are thus similar across these specialties—how to expand their reach. There are several options: train a larger number of lower cost providers in the skill set needed to, at least partially, fill the gap, and serve as an intermediary (e.g., cardiology focused physician assistant), train community members who are willing to commit to staying in the community for free (funded training grants), or use technology to expand access beyond what can be achieved in physical face to face meetings (e.g., phone or video based appointments or virtual consultations). To date, genetic counselors have focused upon the latter solution, but uptake has been slow for both phone and video based genetic counseling appointments. Just recently several startups have begun offering online genetic counseling, though much work is yet to be done. There is also great potential in providing basic genetic counseling content via videos and other mediums. For example, what DNA is and what types of results you might receive from a test (pathogenic, benign, variant of undetermined significance) could easily be conveyed through videos and short educational modules. By decreasing the amount of content a genetic counselor needs to convey in person, and collecting family history on a digital platform prior to the appointment, the time a genetic counselor spends with a patient could be dramatically reduced, increasing genetic counseling capacity. In addition, medically affiliated providers, such as nurses, could be trained to perform genetic counseling for specific conditions, such as drug adverse events or efficacy, hemochromatosis, or other diseases that do not have the same "red flag" level of concern that say cancer or Huntington's disease does.

Taken together these strategies can enhance detection of at risk individuals, promote cascade screening of family members, and expand the reach of genetic and genomic testing.

Genetic screening in the population

Currently medical genetic testing is limited to several scenarios where clinical action is likely. This could be for diagnosis when a clinical presentation suggests a genetic etiology, a prenatal setting where results may change reproductive decisions or early life treatment, or cascade testing in a family.

As the cost of genetic testing decreases, there is growing interest in population level genetic screening. The general principles of population screening are still relevant in the context of genetics. These general principles include recommendations that screening should address a population health priority, should lead to effective treatment or prevention, should have high sensitivity and specificity, and should be cost effective. Some genetics experts advocate that screening for highly penetrant mutations in genes like *BRCA1*, *BRCA2*, and *LDLR* meet all of these criteria. However, polygenic risk scores for complex traits do not currently meet most population screening criteria, though some argue that they may meet these criteria in the future. Still others argue that once testing a few genes meet population screening criteria, full genome screening will become cost effective, and thus society should move toward widespread genomic screening. However, appropriate strategies for analysis of population genome screening will be incredibly complex. As such, full discussion of genetic screening for population health is outside the scope of this book.

Any of these genetic screening advances will lead to a greater need for using family health history in medical care. One misconception about genetic testing is that greater understanding of the human genome and the ability to sequence genomes will decrease the utility of family history. After all, why would a provider need to ask about family cancer history if she has access to the full palette of genetic cancer risk markers. Our belief, and one motivation for writing this book, is that family history will paradoxically become more important in the context of genetic advances, and that its importance will endure for at least the next several decades.

Community screening is likely to increase the importance of family history in internal medicine for several reasons. (1) When genes have large effects, rare variants are best interpreted through family analysis. This is described in detail in Chapter 4. (2) Complex disease is an interplay of genetics and environmental factors. These environmental factors often start in childhood and travel in families. Many genetic effects are only present in certain situations, which may be best ascertained through the family's health history. Thus polygenic risk scores will be most useful with family health history, rather than as a substitute for family health history. (3) Increasingly genetic research will highlight disease risk factors that are not genetic. Heritability of most complex traits is less than 50%, indicating that much of the variability in the population comes from things that are not genetic. By highlighting disease pathways, genetics research has, in some situations, started to shed light on non-genetic risk pathways. As these pathways become clearer, family health history will continue to be a way to identify an important source of risk factors. (4) Family health history will highlight situations where our genetic understanding does not fit with test results and further genetic exploration is warranted. In our clinical practice we see cases every month where family history strongly suggests a genetic syndrome, but initial genetic testing comes back negative. In an important minority of these cases, the persistence of a provider in pursuing genetic reanalysis has identified genetic variants that were unexpected and unidentifiable without extra-ordinary testing and genetic data analysis. Until there is perfect genetic knowledge, providers who collect thoughtful and complete family health histories will continue to contribute knowledge that is a major source of genetic discovery.

Families can help each other understand genetics and prevent disease

Much of this book has been about how to use family health history in the context of medical care. However, one of the most powerful ways to improve health is for families to talk to each other about their own health and the health of other relatives. This awareness is often coupled with warnings from loved ones about behaviors that cause bad outcomes in parents, grandparents, and cousins; as well as nudges to behave better.

Social media technology, like Facebook and Instagram, have created new ways for large families to connect and share information, but it is not clear if these platforms facilitate communication to improve health or impede healthy behaviors.

Genealogy services like Ancestry.com and FamilySearch.org allow family history enthusiasts to connect with distant relatives and pool information in merged pedigrees, occasionally facilitating health discussions. Although focused primarily on documenting family history, these genealogy sites have slowly moved into health. One primary reason for this shift is the rapid growth of direct-to-consumer DNA testing for ancestry. By 2019 over 20 million Americans used AncestryDNA, 23andMe, MyHeritageDNA, or another service to find relatives via DNA matching (Fig. 1).[1,2]

The potential to link family health history with online family history repositories is obvious, but there are technical and privacy issues that still need to be resolved before health systems start downloading family health history from companies like Ancestry.com. On the other hand, the growth of these services is providing new opportunities for distant relatives to connect, and potentially to identify and connect with previously unknown genetic relatives that can enhance health prevention efforts. In fact, this is one of the most promising avenues for adoptees to learn about their genetic risks.

Genealogy and social media may create new avenues for improving the prevention of genetic disease. The founder mutations that cause cancer which were discussed in Chapter 3 each arose from a single ancestor. All other inherited monogenic risk mutations similarly are linked to large or small families. If all living descendants of each founder ancestor knew about their familial connections, through genealogy and social media, they could potentially undergo appropriate genetic testing to know if they should have increased screening or other preventive medicine strategies. Although preventing disease among all the relatives of your patient is far from standard practice today, we describe this

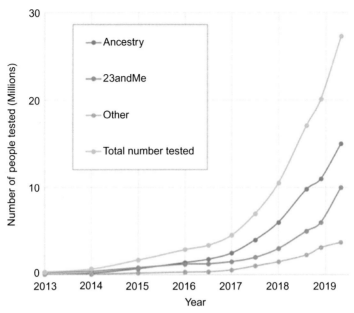

FIG. 1 The increase in direct to consumer testing.[1,2]

concept to illustrate the vast, unexplored potential for the truly ambitious family health historian.

Brian H. Shirts is the lead author for this chapter. First person statements are from his point of view.

References

1. International Society of Genetic Genealogy wiki. *Autosomal DNA Testing Comparison Chart.* International Society of Genetic Genealogy Wiki; 2020. Retrieved from https://isogg.org/wiki/Autosomal_DNA_testing_comparison_chart.
2. Khan R, Mittelman D. Consumer genomics will change your life, whether you get tested or not. *Genome Biol.* 2018;19(1):120. https://doi.org/10.1186/s13059-018-1506-1.

Epilogue

The aspiration for this book has been to facilitate the use of family health history by providers in primary care, especially when managing healthy patients with a positive family health history for a specific chronic disease. It has not been a goal of this book to present an extensive listing of common and not so common chronic diseases which may be revealed through a meticulous or expanded family pedigree. Rather, the goal is to improve the process of recognizing family patterns of disease even when the source of genetic variation may or may not be identifiable at this moment in the history of genomic medicine. Further, it is hoped that this book's contents will spur further work to identify valid and effective strategies for screening families whose history suggests an inherent risk.

The Human Genome Project has been a cornerstone for advancing medicine since its completion at the turn of the century. The collateral medical developments attributable to the project's databases continue to take medical diagnostics and practice to unforeseen frontiers. Some of the insights, however, have spawned questions about how much we truly understand about the mechanisms that underlie disease progression and the possibility that the shared disease risk of genetically related individuals often involve genes and processes that are still unknown or not understood. In the present, this uncertainty and complexity again highlights the importance and utility of family health history for recognizing a familial based disease risk.

The utility of family health history is also relatively simplistic, partly because genetic variants are frequently treated as causative factors, when in fact they usually act at a level that sets off relatively subtle changes that increase vulnerability but do not lead directly to adult-onset disease. In reality, most variants alter the chronic function of cells in which they are active, but at a level far removed from the cellular events most closely tied to disease onset. The risk-elevating impact of pathogenic variants, therefore, is a cumulative effect that is advanced or offset by other gene variants, environmental exposures, and cellular processes. For the provider, the mitigating complexity typically has hampered the development of protocols for managing a healthy patient. Moreover, seeking precise molecular answers for most families with a positive family history may not be feasible. Even where a strong family history is present, genetic test results are often uncertain or negative. In this scenario, it is likely that in the foreseeable future, the collection of family health information will become *more* important, especially as clinical test results and symptoms seen in family members prior to an adverse health event are compiled and compared. This is the kind of clinical information that might be employed to extend productive lifespan, reduce the likelihood and severity of chronic disease and reduce the cost of healthcare for unaffected members of a family with a positive history. These are also the outcomes sought by the most ardent advocates of precision medicine.

Some resources for patients

Each of the URLs for online resources listed here were live when this appendix was created. However, the exact URLs may not be up to date after publication. We encourage readers to visit the websites of organizations that publish these resources, as readers are likely to find similar, updated resources.

Overview of genes, genetics

http://www.learninggenetics.org/

Importance of family health history and worksheets for collecting it

- The CDC has information for patients about the importance of family health history and how to collect it (https://www.cdc.gov/genomics/famhistory/knowing_not_enough.htm).
- The AMA has information for patients about the importance of family health history and how to collect it (https://www.ama-assn.org/delivering-care/precision-medicine/collecting-family-history).
- Kaiser has a video on what to ask relatives about family health history (https://www.youtube.com/watch?v=xSuMkrhzXGA).
- CrashCourse video on heredity (https://www.youtube.com/watch?v=CBezq1fFUEA).
- Mayo clinic video on how to talk with relatives and what to ask about (https://www.youtube.com/watch?v=pIoILk7fKag).
- The Jackson Laboratory has a catalogue of family health history collection tools and resources for patients and physicians (https://www.jax.org/education-and-learning/clinical-and-continuing-education/cancer-resources/family-history-collection).
- The National Library of Medicine has information on the value of family health history as well as the basics of heredity (https://ghr.nlm.nih.gov/primer/inheritance/familyhistory).

Blood clots that run in families/Thrombophilia

Your personal or family history indicates that you or a family member may be at risk for having thrombophilia. Thrombophilia is an inherited or acquired condition that increase a person's risk of developing thrombosis. Thrombosis is also known as abnormal blood clotting.

The National Blood Clot Alliance is a non-profit health organization dedicated to prevention, early diagnosis and treatment of blood clots. They have a number of helpful resources available on their website. They have a fact sheet about the genetics of thrombophilia which you can view and download (https://www.stoptheclot.org/news/the-genetics-of-thrombophilia/).

They also have a fact sheet about how to do genetic or family testing for clotting disorders (https://www.stoptheclot.org/documents/fam_test.pdf).

Breast cancer

Your personal or family history indicates that you or a family member may be at risk for having hereditary breast cancer.

The American Cancer Society provides general information about breast cancer risk, prevention, detection, and diagnosis on their website (https://www.cancer.org/cancer/breast-cancer.html).

The National Human Genome Institute (NHGRI) which is part of the National Institutes of Health (NIH) provides additional information and links to a number of resources about hereditary breast cancer on their website (https://www.genome.gov/10000507/learning-about-breast-cancer/).

Jackson Laboratory, a research institution, has provided more detailed information about hereditary breast and ovarian cancer on their website (https://www.jax.org/education-and-learning/clinical-and-continuing-education/cancer-resources/hereditary-breast-and-ovarian-cancer-syndrome-factsheet).

Breast cancer screening recommendations suggest having annual mammograms starting between the ages 40–45 for women at average risk. Learn more about the screening guidelines and mammograms from the American Cancer Society's website (https://www.cancer.org/cancer/breast-cancer/screening-tests-and-early-detection/mammograms/mammogram-basics.html).

Women at increased risk for breast cancer based on their personal or family history may also have breast MRI scans along with annual mammograms for improved detection of breast cancer (https://www.cancer.org/cancer/breast-cancer/screening-tests-and-early-detection/breast-mri-scans.html).

Cardiovascular disease

Your personal or family history indicates that you or a family member may be at risk for developing cardiovascular disease. There are many conditions that are types of cardiovascular disease. These include heart disease, heart attack, stroke, heart failure, arrhythmia (or abnormal heart rhythm), and heart valve problems. You can learn more about different types of cardiovascular disease from the American Heart Association (https://www.heart.org/en/health-topics/consumer-healthcare/what-is-cardiovascular-disease).

Cardiovascular disease is a multi-factorial condition. This means that many factors including genetics and environment cause the disease. Therefore, cardiovascular disease can run in families. You can learn more about the features that increase risk for cardiovascular disease from the Centers for Disease Control and Prevention (https://www.cdc.gov/heartdisease/family_history.htm).

Knowing your genetic risk of cardiovascular disease can help you take precautions to prevent it.

Colon cancer

Your personal or family history indicates that you or a family member may be at risk for having hereditary colon cancer.

The American Cancer Society provides a basic overview of screening and genetic testing for people with a family history of colon cancer. That information is available on their website (https://www.cancer.org/cancer/colon-rectal-cancer/causes-risks-prevention/genetic-tests-screening-prevention.html).

The National Human Genome Institute (NHGRI) which is part of the National Institutes of Health (NIH) provides additional information and links to a number of resources about hereditary colon cancer on their website (https://www.genome.gov/10000466/learning-about-colon-cancer/).

A colonoscopy may be recommended for patients at risk for developing colon cancer. The American Cancer Society recommends that people at average risk of colorectal cancer start regular screening at age 45. Those at increased risk due to family history may need to start regular screening earlier. You can learn more about colon cancer screening at the American Cancer Society website (https://www.canccr.org/cancer/colon_rectal-cancer/detection-diagnosis-staging/acs-recommendations.html).

One commonly used screening tool is a colonoscopy. You can learn more about colonoscopy at this American Cancer Society website (https://www.cancer.org/treatment/understanding-your-diagnosis/tests/endoscopy/colonoscopy.html).

You may want to speak to a genetic counselor about your risk of hereditary colon cancer. Genetic counselors are healthcare providers with special training in genetics. They talk to individuals with a personal or family history of cancer to help determine if the family cancer is inherited and coordinate genetic testing. You can learn more about genetic counseling for colon cancer by watching this video by the National Society of Genetic Counselors (https://www.youtube.com/watch?v=KlhLyoePnIM).

Diabetes

Your personal or family history indicates that you or a family member may be at risk for developing diabetes.

Diabetes is a multi-factorial condition. This means that many factors including genetics and environment cause the disease. Therefore, diabetes can run in families. You can learn more about the genetics of diabetes from the American Diabetes Association website (https://www.diabetes.org/diabetes/genetics-diabetes).

The Centers for Disease Control and Prevention also has useful information about family history and diabetes on their website (https://www.cdc.gov/genomics/famhistory/famhist_diabetes.htm).

Familial hypercholesterolemia

Your personal or family history indicates that you or a family member may be at risk for having familial hypercholesterolemia (FH), a genetic condition that causes high cholesterol.

The FH Foundation, a non-profit organization which aims to raise awareness of FH provides detailed information about familial hypercholesterolemia on their website (https://thefhfoundation.org/familial-hypercholesterolemia/what-is-familial-hypercholesterolemia).

Additional information is available from the Rare Disease Database (https://rarediseases.org/rare-diseases/familial_hypercholesterolemia/).

You may want to speak to a genetic counselor about your risk of familial hypercholesterolemia. Genetic counselors are healthcare providers with special training in genetics. They talk to individuals who are at risk for genetic disease about diagnosis. They can assess your personal and family history and help determine if genetic testing is appropriate for you or your family members.

Ovarian cancer

Your personal or family history indicates that you or a family member may be at risk for having hereditary ovarian cancer.

Hereditary ovarian cancer may be associated with other types of cancers that run in families including hereditary breast cancer. You can learn more about hereditary breast and ovarian cancer from the Centers for Disease Control and Prevention (https://www.cdc.gov/genomics/disease/breast_ovarian_cancer/breast_cancer.htm). More information about hereditary ovarian cancer is also available at the National Ovarian Cancer Coalition website (http://ovarian.org/about-ovarian-cancer/am-i-at-risk/do-i-have-a-genetic-predisposition).

You may want to speak to a genetic counselor about your risk of hereditary ovarian cancer. Genetic counselors are healthcare providers with special training in genetics. They talk to individuals with a personal or family history of cancer to help determine if the family cancer is inherited and coordinate genetic testing. You can learn more about genetic counseling for ovarian cancer by watching this video by the National Society of Genetic Counselors (https://youtu.be/LPqXf9ecnTY).

Variants of uncertain significance

ClinVar is the best resource for understanding the current state of variant classification for specific variants (https://www.ncbi.nlm.nih.gov/clinvar/).

ClinGen is a consortium of individuals from academic and commercial clinical laboratories that establishes rules for variant classification (https://clinical-genome.org/).

The Human Variome Project is an international organization with projects to help share and understand genetic variation across the world (https://www.humanvariomeproject.org/).

People with specific variants may be interested in registering with studies that connect patients with Variants of Uncertain Significance with researchers, such as the PROMPT study (http://promptstudy.info/).

FindMyVariant has information about how someone can use family information to help classify their familial variant of uncertain significance (https://findmyvariant.org/).

ConnectMyVariant has information about connecting people with the same variant so that they can help other close and distant relatives get genetic testing to help prevent hereditary disease (http://connectmyvariant.org/).

Some resources for doctors

The National Comprehensive Cancer Network develops guidelines for assessing risk of hereditary cancer syndromes and on treatment and management. Access to the guidelines requires an account but you can create an account for free (https://www.nccn.org/professionals/physician_gls/default.aspx).

Familial hypercholersterolemia article reviewing risk assessment options (https://www.ncbi.nlm.nih.gov/pubmed/30382952).

The American College of Cardiology also has expert recommendations for risk assessment (https://www.ncbi.nlm.nih.gov/pubmed/30382952)

A review article on the genetics of cardiovascular diseases (https://www.ncbi.nlm.nih.gov/pmc/articles/PMC3319439/).

The Familial Hypercholesterolemia (FH) Foundation has resources for providers on diagnosis, treatment, and the genetics of FH (https://thefhfoundation.org/familial-hypercholesterolemia/what-is-familial-hypercholesterolemia).

The Jackson Laboratory has a catalogue of family health history collection tools and resources for patients and physicians (https://www.jax.org/education-and-learning/clinical-and-continuing-education/cancer-resources/family-history-collection).

The NIH held a Family Health History tool meeting and produced a document with the descriptions of the tools (https://www.genome.gov/Pages/Health/HealthCareProvidersInfo/FHHT_Workshop_Short_Descriptions.pdf).

Index

Note: Page numbers followed by *f* indicate figures, *t* indicate tables, and *b* indicate boxes.

215

Printed in the United States
By Bookmasters